Flying Jets

Flying Jets

Linda D. Pendleton

Boston, Massachusetts   Burr Ridge, Illinois
Dubuque, Iowa   Madison, Wisconsin   New York, New York
San Francisco, California   St. Louis, Missouri

McGraw-Hill

*A Division of The **McGraw·Hill** Companies*

4 5 6 7 8 9   BKM BKM   9 0 9 8 7 6 5 4 3 2 1 0

Library of Congress Cataloging-in-Publication Data
Pendleton, Linda D.
    Flying jets/by Linda D. Pendleton
       p.   cm.
    Includes index.
    ISBN 0-07-049296-4 (h)
    1. Jet planes--Piloting.   I. Tiltle.
TL711. J38P46   1995
629.132'5249--dc20   95-39231
                                                CIP

Acquisitions editor: Shelley IC. Chevalier
Editorial team: Robert E. Ostrander, Executive Editor
                Norval G. Kennedy, Book Editor
Production team: Katherine G. Brown, Director
                Jan Fisher, Desktop Operator
                Rose McFarland, Desktop Operator
                Linda L. King, Proofreading
                Jodi L. Tyler, Indexer
Design team: Jaclyn J. Boone, Designer   0492964
            Katherine Lukaszewicz, Associate Designer   AV1

To my instructors
John Goodpaster,
David J. Dwyer,
and Stelios Rapis,
You've taught me much!

To my students over the years,
You've taught me the most!

Thanks!

Contents

Foreword

TO ANY PILOT WITH A PISTON-POWERED BACKGROUND, THERE IS NO greater thrill than transitioning to jets. You are flooded with excitement and sensations: the thrill of hearing a jet engine wind up on engine start, so full of promise; the semisweet smell of jet fuel; the exhilaration of hearing jet engines follow you wherever you go; and the power, oh so much power, and all at the command of your right hand.

Along with this excitement comes the pleasant discovery of new concepts, a new flight environment, and elegant systems that provide a new level of comfort and safety.

With an unusual background of jet instruction, Linda Pendleton is uniquely qualified to introduce the wonderful world of jets to the reader. After years of working as a jet instructor for the prestigious FlightSafety International, Linda was chosen by the Federal Aviation Administration for the weighty task of evaluating and licensing new jet pilots as a designated pilot examiner. In this capacity, Linda has examined more than 500 pilots for their first jet rating; as a flight instructor, she introduced the world of jets to many hundreds more. All in all, Linda has more than 10,000 hours as a pilot and 6000 hours of instructing experience.

For many years Linda has been our examiner for our annual pilot proficiency check in our Citation. We are of course always exhausted after Linda's thorough oral and inflight exam, but beyond that, we always have a great time because of Linda's irresistible enthusiasm for flying jets. Best of all, even though we have flown our jet on the line for thousands of hours, Linda always teaches us something new to think about on every flight. In *Flying Jets*, you, too, will benefit from Linda's uncommon experience, special insight, and sense of fun.

In fairness, we must caution you. After you read this book, you will be fired with a passion for flying jets. It'll change your aviation life. So get ready, strap in, and hang on.

John & Martha King
King Schools
San Diego

John and Martha King

Introduction

THIS IS THE BOOK I WANTED WHEN I WAS LEARNING HOW TO FLY JETS. This is the book I have wanted to give to my students who are learning about jet aircraft. I've looked in every aviation bookstore on every airport that I've visited (and I've been to quite a few). I've scanned every catalog I've gotten my hands on, and I've never found the book that I've sought. That is the first reason I wrote this book.

The second reason for preparing this work stems from my philosophy about aviation and aviation education. All too often instructors and students alike believe that serious study of aircraft systems and performance is pertinent only to those pilots pursuing an airline transport pilot certificate, a flight engineer certificate, or a type rating. Many say that this type of material is too advanced for the typical pilot and is better left to the professionals. Nothing could be further from the truth.

First, many FAR Part-23 general aviation aircraft have systems as complicated, and occasionally more complicated, as some found on jets. It is true that most pilot operating handbooks for Part-23 aircraft do not give the in-depth systems diagrams and systems descriptions found in the manuals for Part-25 transport category aircraft, but that does not mean that pilots should not study and understand those systems. It means, however, that the information might be more difficult to find.

Second, professionalism in aviation is not a job title or an airman rating, but is an attitude. I know student pilots with a far more professional approach to their flying than some corporate and airline pilots I am acquainted with. Professionals will always strive to improve their skills and increase their knowledge in all areas of aviation. Nonprofessionals will try to get by with the bare minimum. This book is for professionals.

A FEW JET MYTHS DISPELLED

Several myths are circulating in the aviation community concerning jet aircraft and their flight characteristics. I heard these stories many years ago when I began flying, and I continue to hear them now. Because these myths might be on your own mind, at least to some extent, it seems sensible to consider them before proceeding any further.

Myth 1—Jets are difficult to fly

I heard this myth for the first time before I took my first flying lesson. Jets are no more difficult to fly than any other complex aircraft and are easier to fly than many more-sophisticated propeller aircraft. Of course this does depend upon your point of view. I distinctly remember the time in my aviation career when I was convinced that the amount of time available on a downwind leg was not sufficient time to check carburetor heat, reduce power, extend the first notch of flaps, retrim the aircraft, and announce my position. After all, the runway I was landing on was only 3000 feet long, and I felt that I needed at least 5 miles to accomplish all the necessary tasks!

Jet aircraft lack many characteristics that require special techniques in propeller aircraft. Jet aircraft do not produce a spiral slipstream. There is no "P" factor or torque. There is only one set of levers—throttles or power levers—to adjust. Mixture is automatic. There are no propellers to set and adjust. Performance is predictable and constant.

Jet aircraft are also inherently aerodynamically cleaner than propeller aircraft and because of this can be a bit harder to slow down than propeller aircraft. This characteristic will require more thinking and planning ahead, but this is a mental discipline, not a physical flying skill.

It is true that jet aircraft require much more precision handling. Because of the airspeeds involved, 1 degree of pitch change can make a tremendous difference in a jet. Jets, however, are equipped with much larger and more-precise attitude indicators (called ADIs or attitude deviation indicators) than are found in other aircraft. (I prefer to think that these are provided to allow us to exercise more precise control of our aircraft, not because by the time most of us reach the left seat of a jet we need to have things in slightly larger "print"!) A pilot who is current on instruments and possesses a good scan and the ability to handle an aircraft precisely will have no problems flying jets. A heavy hand on the controls is counterproductive in any aircraft. Jets only insist that you refine your aircraft handling. Many propeller aircraft I have flown require at least the precision attitude control required for high-speed, aerodynamically clean jets.

Myth 2—Jets are unrecoverable if flown anywhere near a stall

I certainly hope this is not true since this is almost the condition that the jet is in every time I land it! Some jet aircraft, particularly those with T-tails, swept wings, and aft-mounted engines, can be unrecoverable if allowed to progress to a deep stall. These aircraft are usually equipped with stick shakers to warn of an approaching stall and stick pushers to avoid any stall. Other jet aircraft can be recovered from deep stalls as readily as any training aircraft.

These handling characteristics should not be a source of apprehension, but should motivate the pilot to refine skills and understand the handling characteristics and limitations of the aircraft flown. Any aircraft (any piece of complex machinery) has the potential for harm if limitations are not respected. I have always maintained that there are no dangerous aircraft, only dangerous and complacent pilots.

Myth 3—Jets all have a very narrow "coffin corner"

The term *coffin corner* describes the area of the performance envelope where the difference between low-speed stall buffet and high-speed Mach buffet is only a few knots indicated airspeed. As you probably remember from your basic aerodynamics, stall speed varies with indicated airspeed (or equivalent airspeed, in some cases). Mach number varies with true airspeed. As the aircraft climbs, the indicated airspeed representing the speed of sound drops; therefore, the indicated airspeed for Mach-buffet speed and the indicated airspeed for a stall come closer together as altitude increases. No civilian jets have enough thrust to climb to the altitudes where this margin is close enough to be of concern.

At 41,000 feet, most civilian jets will have at least a 100- to 150-knot margin between low-speed buffet and high-speed Mach buffet. At 51,000 feet (rarefied air to be sure), this margin might drop to 45 knots, which is still plenty of airspeed because aircraft certified at those altitudes usually require the use of the autopilot to fly at such lofty altitudes. Figure I-1 shows the relationship between low-speed buffet speed and Mach-buffet speed for a Lear 36 from 10,000 feet to Flight Level 450 (45,000 feet).

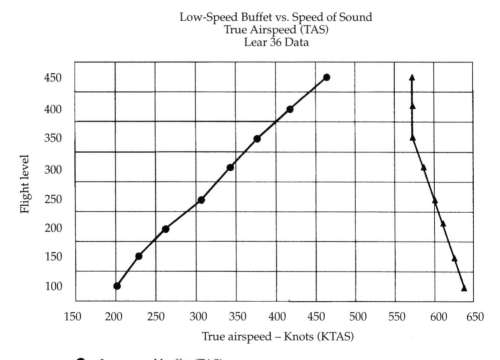

Low-Speed Buffet vs. Speed of Sound
True Airspeed (TAS)
Lear 36 Data

● Low-speed buffet (TAS)

▲ Speed of sound (TAS)

Fig. I-1. The true airspeed for onset of low-speed buffet and the true airspeed for the speed of sound is plotted here for the Lear 36 aircraft. The area where the two lines begin to converge is known as the "coffin corner."

ORGANIZATION AND MATERIAL COVERED

We will begin our exploration of jet aviation with a review of basic aerodynamics. Many pilots study this subject only to pass written tests and forget or ignore the practical everyday application of the principles of aerodynamics. Understanding the forces acting upon an aircraft in flight goes beyond merely reciting the names of the four forces that the FAA insists you know to pass written tests. With understanding comes mastery, so we will spend some time in review.

Lift and drag will be discussed with a view toward the application of these principles to jet aircraft. The relationship between induced and parasite drag will be discussed along with their effects on aircraft performance. Weight and center of gravity will be applied to aircraft stability and control. The methods of determining maximum limits of CG during certification will be explained. The effects of aircraft loading on performance will be explored.

In chapter 2, high-speed and high-altitude aerodynamics will be covered. Mach number, subsonic, transonic, and supersonic speed ranges will be defined. Compressibility, shock waves, and supersonic flow patterns will be discussed and illustrated. The effects of configuration and airfoil sections and control surfaces will be explained along with stability and control. Dutch rolls and yaw dampers plus stick shakers, pushers, and pullers will be discussed and illustrated.

Chapter 3 introduces a subject all too often given only passing attention by students and instructors alike. High-altitude physiology is a crucial area of knowledge that is central to the continued health and safety of all who fly. Perhaps the disinclination to discuss topics of potential serious risk is inherent in humans; however, this attitude is counter to safety issues. All humans will be hypoxic to a greater or lesser degree when operating at any altitude above that of normal residence, but few humans will admit the fact.

Chapter 4 begins with a short course in physics to establish and define the terms used to discuss the design, theory, and operation of gas turbine—jet—engines. The main sections of a typical jet engine will be defined and illustrated, and each section's importance to the operation of the engine will be explained. Although turbine engines are considered one of the things that make jets so "different" from piston aircraft, we will discover that all engines propel aircraft by using the same mechanism, the acceleration of air.

The chapter will continue with a discussion of the supporting systems of a jet engine and the engine and aircraft systems that use engine bleed air for their operation. As stated, while all aircraft engines are essentially air pumps, a jet engine accelerates much more air than is necessary for aircraft propulsion; therefore, the bonus air can be used for other tasks. This is a major difference between piston and gas turbine engines. The use and abuse of thrust reversing systems will be presented.

To conclude chapter 4, the major operating characteristics and instrumentation of jet engines will be explored. Effects of altitude, compressibility, ram recovery, and pressure ratios will be explained and explored. The efficiency of jets at altitude and the effect of speed and altitude on propulsive efficiency will be discussed.

Chapter 5 will introduce the reader to jet-aircraft environmental systems. A thorough discussion of aircraft pressurization, which will be generic to almost all pressurized aircraft, will be undertaken. Air-cycle machine, air-conditioning pack, and temperature control systems will be explained, as will supplemental Freon air conditioning that is found on some aircraft. A presentation of supplemental oxygen systems will conclude the chapter.

In chapter 6, some of the other systems found on high-performance turbine aircraft will be illustrated. Antiskid and power-braking systems that have been common on large jet aircraft since the early 1960s are now being touted as the "newest technology" on automobiles. The workings and benefits of these systems will be discussed. Spoiler and speed-brake similarities and differences will be explored along with other lift-modifying devices such as leading-edge devices. Nosewheel steering systems and their control mechanisms will be explained.

Chapter 7 will provide an overview of the various flight-guidance systems found on turbine-powered aircraft. The various modes and programming options for flight directors and autopilots will be illustrated along with the navigational inputs to these systems from the various available systems in common use.

Chapter 8 will discuss the actual flying and flying characteristics of jet aircraft. The regulatory requirements for performance certification and the calculations that result from these regulations will be thoroughly explained. Some misunderstandings and false confidences in performance figures will be exposed along with their reliability and necessity.

The events and calculations of an actual flight from Van Nuys, California, to Las Vegas, Nevada, will be followed. Differences between a jet's takeoff, climb, and en route profiles will be pointed out, as will the different handling characteristics of the aircraft. The advantages realized with the lack of "P" factor and drag associated with propellers plus the disadvantage of a lack of the gyroscopic stabilizing force of propellers will be discussed.

The methods and procedures used in executing the various types of instrument approaches will be explained, as will missed approaches, go-arounds, and balked landings. The exploration includes maneuvers found on the ATP or type-rating practical tests, such as steep turns, approaches to stalls, and emergency procedures.

Finally, chapter 9 will set forth the principles and practices of cockpit resource management and crew coordination. The concepts of information sharing, situational awareness, and decision making in a coordinated crew environment will be explained along with the underlying necessity for these procedures. Pilot incapacitation and its recognition and remedies will be set forth along with measures to prevent its occurrence. The chapter concludes with a discussion of professionalism and its importance in any aircraft cockpit, be it single-engine trainer or jumbo transport.

The glossary and index should help the reader when defining, locating, and cross-referencing key words and concepts.

The reader will note that many systems found on jet aircraft have not been covered. These systems, such as electrical systems, hydraulic systems, and fuel storage and delivery systems, are found on almost every aircraft and do not, in themselves, differentiate jet aircraft from others. In most cases, the difference between these systems as found on jets and the similar systems found on smaller or piston-powered aircraft is more of a size difference rather than a complexity difference.

When viewed from an operational standpoint, these systems are generic in their nature, and the differences in their design and engineering do not affect the flying qualities of the aircraft. Landing gear is, after all, landing gear, and trailing-edge wing flaps are trailing-edge wing flaps, and the method of their deployment or retraction does not impact the flying characteristics of the aircraft. True, the differences in kinds of flaps (split, Fowler, etc.) do make a difference in flying characteristics, but the power mechanism for deployment does not.

It can be said that the electrical systems found on large jet-transport aircraft differ greatly from those found on smaller jets and piston-powered aircraft, but again, these differences are mainly in design and engineering. From an operational standpoint, it matters little how the electricity is generated or whether it is alternating or direct current. The pilot is generally interested in having the electrical energy that is necessary to power essential instruments or the procedures for operation if the energy source fails.

You also will not find a large number of mathematical formulas in this work. The formulas I have included are presented to illustrate a point, not as items of memorization. Thankfully, the training world, and especially the jet training world, seems to be moving away from the old military style of teaching and rote memorization of volumes of highly technical and highly detailed information, however useless that information might be. An understanding of concepts and a correlation of concepts learned is far more important to the safety of flight than knowing that exact lift equation or the number of holes in the speed brake of an aircraft.

Aircraft systems should be learned from the operational standpoint: "What do I need to know to safely operate this aircraft, and how do I solve the problems presented by system malfunctions?" Is it more important to know the exact pressure at which a FUEL PRESSURE LOW or an OIL PRESSURE LOW light illuminates, or the proper procedure to be taken to safely complete the flight?

I know some will argue that I am oversimplifying the process of jet type-rating training, but is this necessarily true or are those old traditionalists simply trying to perpetuate the myth that a woman or man must be superhuman to fly jets? There is always material available for the airplane junkie (I confess, I belong to this group!) to learn everything available about the aircraft in question in great detail, and I would not discourage that learning—after the aircraft has been mastered from an operational standpoint.

1

A review of basic aerodynamics

ALL PILOTS HAVE BEEN TAUGHT SINCE THEIR FIRST HOURS OF FLIGHT instruction about the four forces acting on an aircraft in flight: lift, weight, thrust, and drag. The interaction and interdependence among these four forces, however, are not usually adequately discussed in primary training.

AIRFOIL TERMINOLOGY

Because the shape of any airfoil and its inclination in the airstream are critical in determining the pressure distribution about the airfoil and subsequently the lift generated, or lack thereof, using the proper terminology is necessary. Figure 1-1 shows a typical cambered airfoil and illustrates the various items of airfoil terminology.

An airfoil, or more properly an airfoil section, is a slice of a wing taken with a plane oriented parallel to the longitudinal and vertical axes of the aircraft. Wing planform (shape as seen from above) sweep back, wingtip effects, taper, and other design features of wings are not taken into consideration when discussing airfoils. The shape of the airfoil and its orientation to the relative wind will determine the lift produced by that airfoil.

NACA airfoils

Many different airfoil shapes exist. Slight variations in airfoil shapes will alter their characteristics in one way or another, and with the development of more different designs, it became necessary to develop a standard means of describing an airfoil shape. Around 1932 the National Advisory Committee for Aeronautics (NACA) tested a series of airfoils that became known as the NACA four-digit series. The geometry of these airfoils was described by four digits, as the name implies. The first digit gives the maximum camber of the airfoil expressed in percent of chord; the second is the location of maximum camber in tenths of the chord length; and the last two are the maximum thickness in percent of chord. For example, the 2412 airfoil is a 12-percent-thick airfoil having a 2-percent camber located 4/10 chord from the leading edge.

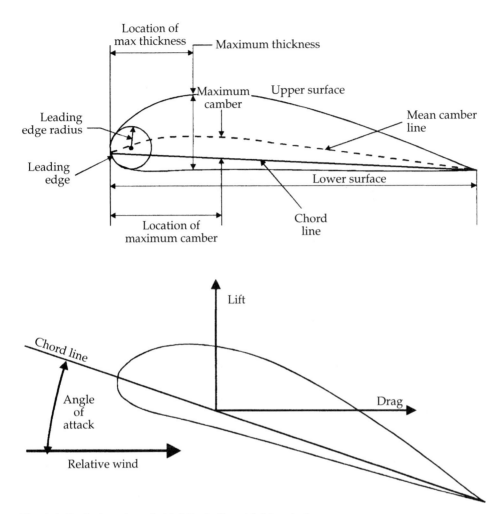

Fig. 1-1. Typical cambered airfoil illustrating airfoil terminology.

The NACA five-digit series of airfoil shapes, first developed around 1935, uses the same thickness distribution as the four-digit series. Mean camber line is defined differently, however, to move the position of maximum camber forward. This is done in an effort to increase the maximum section lift coefficient. The numbering system for the five-digit series is not as straightforward as that for the four-digit series and requires a bit of math to learn the true picture. The first digit multiplied by 3/2 gives the design lift coefficient (C_L) of the airfoil. The next two digits are twice the position of maximum camber in percent of chord. The last two digits give the percent of thickness in the same presentation as the four-digit series. For example, the 23012[1] airfoil is a 12-percent-thick airfoil having a design C_L of 0.3 and a maximum camber placed 15 percent of the chord back from the leading edge.

The NACA 1 (Series 16) airfoils followed in around 1939 and these were the first wing sections to be based on theoretical considerations. These airfoils have seen extensive use in both marine and aviation propellers due to their unique characteristics of avoiding low-pressure peaks in areas of high velocity and delaying the onset of adverse effects from shock waves in regions of high velocities. The 1-Series airfoils are also identified by five digits, for example NACA 16-212. The first digit designates the series. The second designates the location of minimum pressure in tenths of the chord. After the dash, the first number gives the section lift coefficient in tenths and, as in other airfoils, the last two digits represent the maximum thickness in percent of chord.

The NACA 6-Series[2] airfoils were designed to achieve desirable drag, compressibility, and maximum lift coefficient performance—a somewhat conflicting set of goals. It does appear, however, that the motivation for these airfoils was primarily the achievement of low drag. The design of these wing sections is conducive to maintaining extensive laminar flow over the leading portions of the airfoil over a limited range of lift coefficient values. Outside these values, the drag coefficient and maximum lift coefficient are similar to other airfoils. The numbering system for the 6-Series airfoils has many perturbations, but the later series of numbers follows the general pattern of the NACA 65_1–212 airfoil where the first digit identifies the series and the second digit is the location of the minimum pressure in tenths of the chord. The subscript numeral 1 indicates that low drag is maintained at section lift coefficients of 0.1 above and below the design C_L of 0.2 denoted by the 2 following the dash. As in other airfoil series, the last two digits specify the percentage thickness of the wing section.

Systematically numbered series of airfoils have given way, at least in part, to specialized airfoils designed to satisfy specific requirements. These airfoils are developed with the use of sophisticated computer programs not available in the early days of airfoil development. This is an appropriate time to discuss some characteristics that vary with airfoil shape.

It has been said that given enough thrust, even a barn door can be made to fly. Actually this statement is not far from wrong. Almost any relatively flat surface will generate lift. Look at the wings of one of the small, hand-launched balsa-wood gliders available in toy departments. These aircraft have flat wings, and yet they generate lift and fly. The flat plate is probably the simplest of airfoils. It is a symmetrical airfoil and therefore needs to be at a positive angle of attack to generate lift.

LIFT

Pressure differences

The lift produced by an airfoil is dependent upon the distribution of air pressure around that airfoil. The pressure distribution is in turn, dependent upon airfoil shape. The barn door has a very simple shape—flat—and generation of lift by this type airfoil depends totally upon the angle of attack. At a zero angle of attack, the airflow over that top and bottom surfaces of the airfoil will be the same, and consequently no lift will be produced. The same is true for a symmetrically curved airfoil. At a zero angle of attack, this airfoil produces no lift[3]. Figure 1-2 shows the air-flow around both a flat-plate and a symmetrical airfoil at zero angle of attack and at a positive angle of attack.

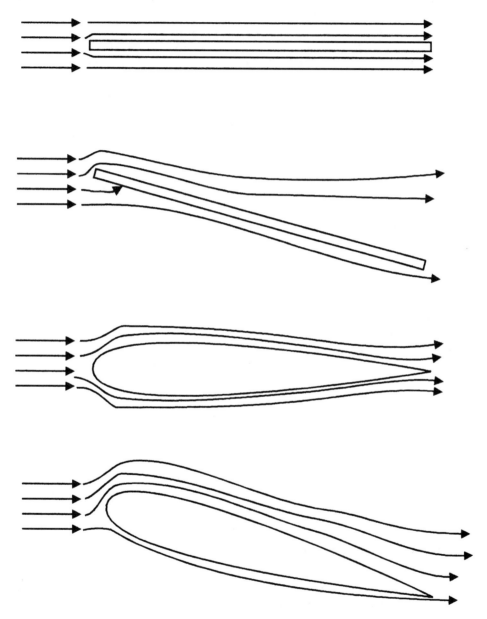

Fig. 1-2. Flat-plate and symmetrical airfoils at zero and positive angles of attack. Notice that there is no difference in airflow, and consequently no difference in air pressure between the top and bottom surfaces of either airfoil at zero angle of attack.

Everyone who has ever been in an aviation ground school has heard of Bernoulli's theorem[4]. This theorem has been stated and misstated, reduced to an equation and expanded to occupy volumes, but simply stated it says thus:

> The total of the static pressure of a parcel of air and the dynamic pressure of that parcel will always be constant. If the static pressure increases, the dynamic pressure must decrease to maintain the constant. Conversely, if the dynamic pressure increases, the static must decrease. If the area through which a parcel of air must move is constricted, the dynamic pressure will increase and the static pressure will decrease.

Now, what do "static pressure" and "dynamic pressure" really mean? The static pressure of a mass of stationary air in a container is the pressure that it would exert on the walls of the container. If air is in motion, it can and will still have static pressure associated with it. Dynamic pressure is that pressure that would be exerted if the air were brought to rest by a dam or barrier. It is that pressure that would be exerted on the barrier and can be thought of as the pressure of motion.[5]

The free airstream in front of an airfoil is considered to have a uniform velocity associated with it across the entire stream. As the wing moves through the air (or the air moves past the wing if we are talking about a wind tunnel) and the airstream approaches the front of the wing, it must divide with part of the airstream flowing above the wing and the rest flowing beneath the wing. If we consider the whole cross-sectional area of the airstream, however, we will see that it is reduced by the amount taken up by the airfoil. (See Fig. 1-3.) According to Bernoulli, this reduction in cross-sectional area will cause an increase in velocity. Because the total energy of the airstream is a constant, this increase in velocity (dynamic pressure) will cause a corresponding decrease in static pressure over the top of the airfoil. There is also a decrease in static pressure (from ambient pressure of the free airstream) on the bottom surface of the wing, but the pressure decrease on the upper surface of the wing is greater resulting in a net force in the upward (actually top of the wing) direction.

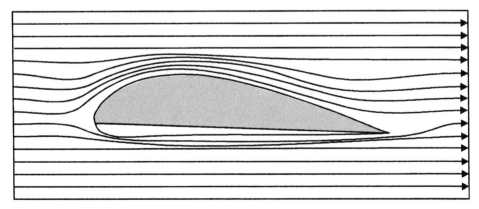

Fig. 1-3. An airfoil in an airflow. It can be seen that the airfoil takes up space in the cross section of the airflow.

At a point on the airfoil near the maximum thickness, maximum velocity of the airstream and minimum static pressure will occur. Because air has viscosity (stiffness, or resistance to flow caused by internal friction of a liquid), some of its energy will be lost to friction and a wake of turbulent, low-velocity air will be found near the trailing edge of the airfoil. This results in a small area of high pressure in that region of the airfoil. (This will be discussed more thoroughly with the discussion of laminar flow.)

Pressure distribution around an airfoil

We have talked about the pressure on the top and bottom surfaces of an airfoil as though those pressures were constant over these surfaces. In fact, the pressures on an airfoil surface will vary dependent upon the position along the chord of the airfoil where the measurement is taken. The significance of the pressures along the surface of an airfoil is the difference between these pressures and the pressure of the free airstream or ambient pressure. Creating a pressure lower or higher than ambient pressure creates an aerodynamic force by virtue of the pressure difference.

Figure 1-4 shows a cambered airfoil at an angle of attack at which it generates zero net lift. Notice that this angle of attack is negative. A symmetrical airfoil would generate zero net lift at a zero angle of attack, but a cambered airfoil will generate lift even at a zero angle of attack. The arrows pointing away from the surface of the wing section show the direction and size of the pressure lower than the surrounding air pressure, also known as negative pressure. Arrows pointing toward the surface show an area of pressure on the airfoil that is greater than ambient and is called *positive pressure*. In a zero-lift condition, the sums of all the pressures pulling the airfoil upwards exactly equal those pulling the airfoil in a downward direction, and the net force acting on the airfoil is zero; therefore, the net lift produced is zero.

When this airfoil is rotated to a positive angle of attack, the pressure distribution will change to that shown in Fig. 1-5. Notice that the size of the negative pressure arrows on the top surface of the airfoil has increased significantly while the size of the arrows on the bottom surface has decreased. This results in a net force

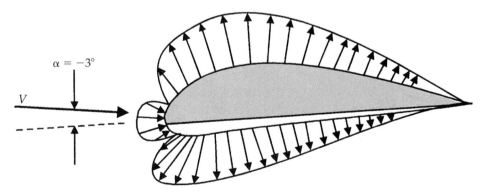

$\alpha = -3°$

V

Fig. 1-4. A cambered airfoil at an angle of attack producing zero lift. Notice that this is a negative angle of attack.

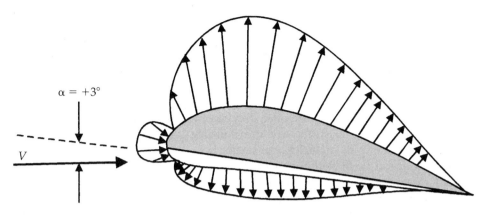

Fig. 1-5. A cambered airfoil at a positive angle of attack producing lift. This angle of attack is typical of a cruising condition.

in the upward direction known as *positive lift*. Also note that the lift generated in a downward direction is still present on the bottom surface of the airfoil, but the net effect is an upward force. This angle of attack would be typical of that found in a cruising condition.

A wing at a very high angle of attack is shown in Fig. 1-6. Notice that not only has the negative pressure on the top surface of the airfoil been greatly increased (or re-

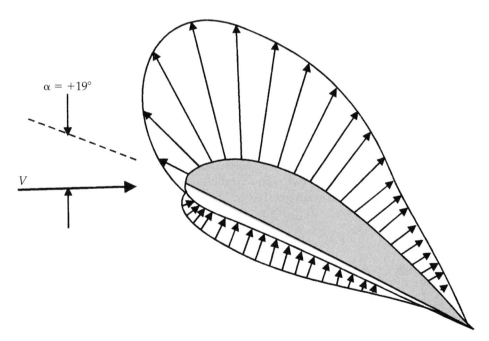

Fig. 1-6. A cambered airfoil at a high angle of attack. At this angle of attack, the pressure on the bottom surface of the airfoil has changed from negative to positive.

duced, depending upon how you look at it), but there is now positive pressure against most of the lower surface of the airfoil, which contributes to the total lift of the airfoil. Also notice that this positive pressure against the lower surface of the wing occurs only at high angles of attack. It is not normally positive pressure that "holds" the airplane in the air, but rather the difference in pressure between the wing surfaces and the surrounding air. Although there is positive pressure against the lower surface of the wing, this pressure is only a minor contributor to the total lift. The pressure distribution along the surfaces of an airfoil changes with the angle of attack.

The forces acting on an airfoil can be mathematically combined to arrive at a point on the airfoil at which these forces effectively act. This is much the same as the mathematics done to arrive at a center of gravity through which the weight of the aircraft is deemed to act. The point on the surface of the airfoil at which these forces are centered is called the *center of pressure*. As can be seen from the illustrations, the center of pressure of an airfoil surface will vary with the angle of attack, and so the resultant force acts at a different point on the airfoil surface for each angle of attack. The center of pressure will move forward with increasing angles of attack.

The differing centers of pressure on an airfoil create an aerodynamic pitching moment. Notice that the pressure distribution along the upper surface of an airfoil differs from that along the lower surface. (This is not so for symmetrical airfoils.) The center of pressure of the upper surface of the wing remains behind that of the lower surface and consequently gives the wing a nosedown pitching moment (Figs. 1-7 and 1-8).

Another contribution

Sooner or later a hangar-flying session will turn to the discussion of the "true" explanation of lift. There will always be those who maintain that Bernoulli was full of hot air and that the lift produced by a wing can be explained totally by analyzing the change of momentum of a mass of air.[6] It is true that as the airflow leaves the trailing edge of the wing, the wing changes the momentum of the air and gives it a downward component. The wing pushes the air downward, and the air pushes the wing upward. This does not, however, run counter to Bernoulli's theorem. It is actually the pressure differential imparted by the shape of the wing that causes the downward acceleration of the air mass. The two methods of explaining lift are not opposing theories, but are only two different methods of explaining the same phenomenon.[7] They are mutually supportive and parts of the same overall picture.

AERODYNAMIC FORCE

The aerodynamic force acting on a wing is the resultant of all the static pressures acting upon an airfoil in an airflow multiplied by the wing area affected by the pressure. The line representing the aerodynamic force vector passes through the chord line at the center of pressure of the airfoil. This force will act in an upward direction perpendicular to the relative wind.

Because lift is the force that operates opposite to weight, and because weight is the force of gravity acting upon the mass of the airplane, it follows that lift acts

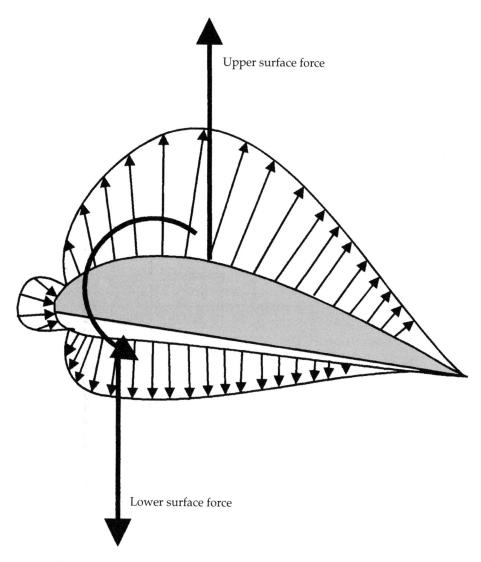

Upper surface force

Lower surface force

Fig. 1-7. A cambered airfoil showing the center of pressures for both the upper and lower surfaces of the wing and the consequent pitching moment.

perpendicular to the ground and not to the relative wind. As can be seen from Fig. 1-9, the magnitude, or length, of the lift component is less than that of the total aerodynamic force vector. If we draw a line perpendicular to the lift component and extend it to a point directly beneath the end of the aerodynamic force vector, we will have another component quantity. This along with lift are the components of the total aerodynamic force vector.[8] This component of the total aerodynamic force is drag, and in particular, induced drag. More about this later.

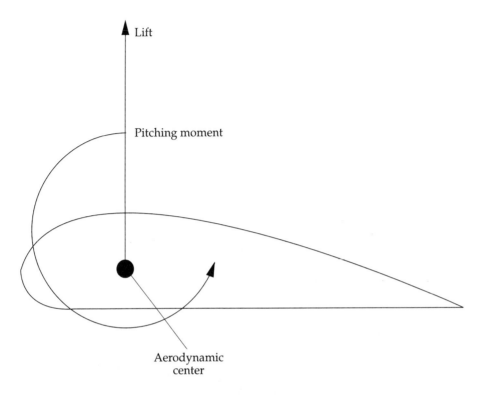

Fig. 1-8. The lift of an airfoil and the pitching moment can be considered to act about the aerodynamic center of the airfoil.

Aerodynamic center

For cambered airfoils, it has been shown that the center of pressure moves forward along the chord line as the angle of attack increases. Conversely, as the angle of attack decreases, the center of pressure moves aft. This movement makes calculations involving stability and stress analysis extremely difficult.

There is, however, a point along the chord of the wing section about which the pitching moment is constant with changing angle of attack while the velocity remains constant. This point is the *aerodynamic center* of the airfoil. The location of the aerodynamic center varies slightly dependent upon the airfoil shape, but subsonically it is located between 23 and 27 percent of the chord aft of the leading edge. Supersonically, the aerodynamic center shifts aft to approximately 50 percent of the chord. The aerodynamic center, unlike that center of pressure, does not change with the angle of attack; thus, if we consider the lift and drag forces as acting at the aerodynamic center, the calculations will be greatly simplified.

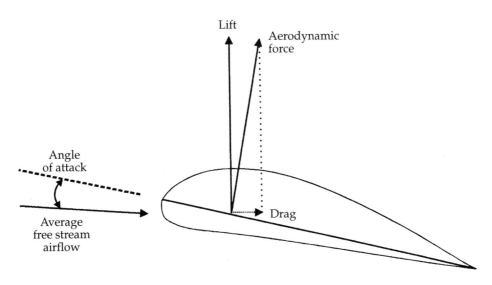

Fig. 1-9. The aerodynamic force on an airfoil is a vector that can be separated into the lift and drag components acting through the aerodynamic center.

THE STALL

As we have seen, as the angle of attack of an airfoil increases, the net lift generated by that airfoil increases. This increase cannot go on indefinitely, however, and for every airfoil there will be an angle of attack at which maximum lift is produced. Beyond this point, the airfoil is incapable of producing the lift required to support the weight of the aircraft and is said to be stalled. Each airfoil shape has an angle of attack at which it stalls and this angle remains constant throughout the flight envelope. Stall speeds might vary, but stalling angle of attack does not.[9]

Several things contribute to the stall. As the angle of attack increases, the stagnation point moves farther down on the forward part of the airfoil making the upper surface effectively longer. As air flows over the surface of the airfoil, it encounters a certain amount of resistance from friction. As the distance traveled increases, the total force of friction increases.

Secondly, there is an effect resulting from the pressure gradient over the wing surface. As air flows over the surface of the wing from the leading edge, it initially encounters decreasing pressure with distance. This is known as a favorable pressure gradient because it tends to help the air to flow along the surface of the wing. As the airflow passes the peak of negative pressure, it encounters a reversal in pressure. Pressure now increases (or becomes less negative) with distance traveled. This condition works against the airflow and is termed an unfavorable pressure gradient. As the angle of attack increases toward the stalling angle of attack, the center of pressure moves forward, and this unfavorable pressure gradient becomes longer and steeper.

Finally, the combined effects of the surface friction and the unfavorable pressure gradient becomes greater than the energy available in the airflow to overcome them. At this point, the airflow will detach itself from the surface of the airfoil. With no airflow over the surface of the wing, there is no longer a reduction of pressure over the upper wing surface, and lift is drastically reduced. In the place of the smooth lift-producing airflow over the wing, there is now a wake of random, turbulent air. Lift is not reduced to zero, however, because at this angle of attack there is normally positive pressure being exerted against the lower surface of the wing. Lift is, however, reduced below that necessary to sustain flight, and gravity will prevail.

Different airfoils have different stalling characteristics. One factor is the roundness of the leading edge; those airfoils having the sharpest leading edges produce the most abrupt stalls. A very sharp leading edge can act as a barrier to airflow at very high angles of attack and cause the airflow to separate at the leading edge. While this can be a very undesirable characteristic, it can be used to advantage when it is desirable to induce a stall at one particular point on a wingspan at an angle of attack lower than the normal stalling angle. The stall is induced by affixing a sharp leading edge device known as a *stall strip* to the leading edge of the wing at a predetermined spanwise point. This will cause the airflow to separate and stall a portion of the wing while the rest of the wing still produces lift.[10]

WINGTIP VORTICES

Until now, we have discussed the wing as though it were a two-dimensional section or profile of a wing. An actual wing, however, has three dimensions and the wingspan and the shape as viewed from above (the planform of the wing) also affect the aerodynamic characteristics of the wing.

An airfoil creates a pressure differential between its upper and lower surfaces. Air will attempt to flow from an area of higher pressure to an area of lower pressure. Along the span of the wing, this tendency results in a net force being exerted on the wing. At the wingtips, however, the pressure differential still exists, but there is no more wing against which to exert pressure (block the flow). The result is a flow around the wingtip from the higher pressure area below the wing to the area of lower pressure above the wing. This flow is known as a *wingtip vortex.*

Wingtip vortices trail behind the wing as the aircraft moves forward. Because the vortex is formed by the pressure differential between the bottom and the top of the wing, it follows that the vortices will be strongest when the pressure differential is the greatest. This will occur at high angles of attack. The more lift the wing is generating, the greater the pressure differential; the slower and heavier the aircraft, the greater the angle of attack.

All aircraft produce wingtip vortices, but those produced by Cessna 152s and Citabrias are not normally as feared as those termed *wake turbulence* and trailed by 757s and DC-10s. Furthermore, the vortices affect not only following aircraft but also the aircraft producing the vortices. The effect on the producing aircraft is not dangerous, but it does reduce the efficiency of the wing.

Downwash effects

As can be seen from Fig. 1-10, wingtip vortices exert a downward motion to the air leaving the trailing edge of the wing. This downward motion has the effect of tilt-

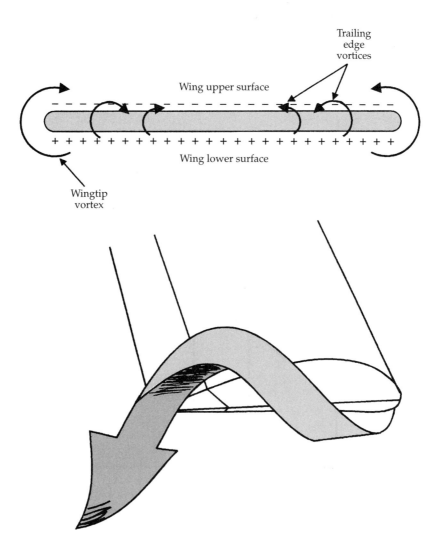

Fig. 1-10. A wing produces a pressure differential between the top and bottom surfaces. This pressure differential causes high-pressure air from the bottom of the wing to flow over the top of the wing and then to swirl downward creating wingtip vortices.

ing the lift vector (aerodynamic force) aft. Now, since not all of the lift is acting in a direction directly opposite the force of gravity, the angle of attack will have to be increased slightly to compensate for the loss of lift (and increase in drag—any force that is opposite the direction of flight is drag) when downwash is present. This additional angle of attack is known as the *induced angle of attack* because it is necessary due to the flow induced by the downwash.

Figure 1-11 shows the lift produced by a hypothetical airfoil at respective angles of attack with and without downwash present. Notice that to obtain the same amount of lift with downwash as could be obtained without it, a higher angle of attack must be maintained. The difference in the angles of attack with and without downwash is the induced angle of attack.

The only way to eliminate the downwash is to eliminate the top vortices. This is an impossibility since only an infinite wing has no tip vortices. The longer the

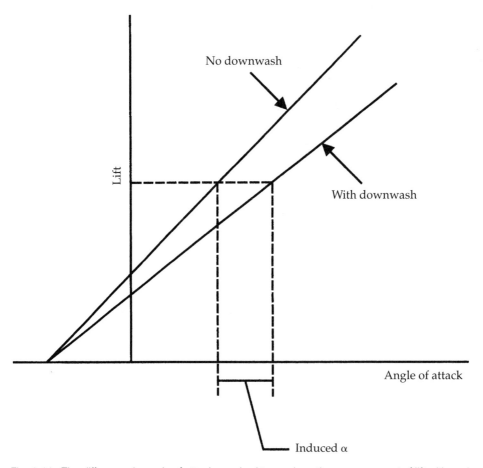

Fig. 1-11. The difference in angle of attack required to produce the same amount of lift with and without downwash is referred to as induced angle of attack.

wingspan is, however, the less effect the wingtip vortices will have on angle of attack. A longer wingspan reduces the angle of attack required for a given amount of lift and consequently reduces the drag associated with higher angles of attack and makes the wing more efficient. One development in this area has been the addition of winglets on the wingtips of newer transport and executive aircraft. The winglets act to effectively increase the wingspan of the aircraft and reduce the drag associated with lift production.

ASPECT RATIO

Increasing the span of the wing without modifying the chord will increase the amount of lift the wing is capable of producing by increasing the wing area. This would also increase the weight of the aircraft. In order to maintain the same wing area, the chord must be reduced proportionately when the span is lengthened. The *aspect ratio* of the wing, or the span divided by the average chord, and its effect on induced drag is the principal effect of the wing planform. With tapered or rounded planform wings, where the average chord might not be readily determined, aspect ratio can also be determined by dividing the square of the span by total wing area. The two equations for calculating aspect ratio are:

$$AR = \frac{b}{c_{av}}$$

$$AR = \frac{b^2}{S}$$

where: b = span
c = chord
S = wing area

Induced drag will vary inversely to wing aspect ratio. In other words, when the wing aspect ratio is increased, the induced drag will decrease. This is because aspect ratio varies directly with wingspan, and increasing span decreases induced drag from tip vortices. Increasing the span will increase the aspect ratio. Figure 1-12 shows two wings with different aspect ratios but the same area. Wing A will suffer more from induced drag in high-lift conditions than will Wing B.

Lift coefficient

Dynamic pressure possessed by a moving fluid is equal to ½ the density (ρ, rho) of the fluid times the velocity squared:

Pressure = $\frac{1}{2} \rho V^2$

A *force* is defined as a pressure times the area over which the pressure acts. The amount of lift obtained from a wing should therefore be the product of the pressure and the area of the wing. Lift is not exactly equal to the product of these two quantities, however, since not all of the force is translated into lift. The portion of the force translated into lift is expressed by the lift coefficient, C_L. Therefore, we can calculate the amount of lift (force), L, that a wing will produce if we know the lift coefficient of the airfoil, the density and velocity of the air, and the area of the wing.

$$L = C_L \times (\tfrac{1}{2} \rho \, V^2) \times S$$

Conversely, we can calculate the lift coefficient if we know the other values:

$$C_L = \frac{L}{\tfrac{1}{2} \rho \, V^2 S}$$

These results are different for each angle of attack since the lift produced by the wing is dependent, among other things, upon angle of attack. At higher angles of attack, we would obtain a higher C_L. If the lift coefficients are calculated and plotted for increasing angles of attack, a graph similar to the one shown in Fig. 1-13[11] will result. Some key characteristics can be determined by studying this type of graph. First of all, note that the maximum lift coefficient, or C_{LMAX}, for this airfoil is 1.8—the peak of the curve. Notice, also that this occurs at an angle of attack, α, of 18°. This indicates that this airfoil will produce more lift than one with a 1.4 lift coefficient. A clue to the abruptness of the stall can be inferred by noting the sharpness of the curve at the peak. A sharp peak indicates an abrupt stall while a more rounded curve will indicate a gentler stall. It can also be deduced that this is a cambered, not a symmetrical, airfoil since lift is produced at a zero α. Zero lift does not occur until approximately –2°. A symmetrical airfoil would produce zero lift at zero angle of attack.

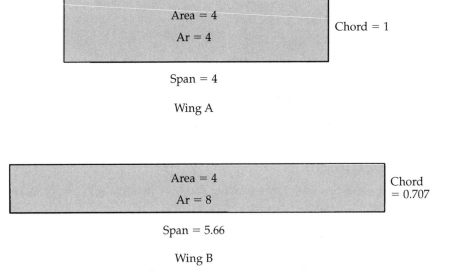

Fig. 1-12. Wing A and Wing B have the same area but different aspect ratios. Wing A will suffer more from induced drag than Wing B.

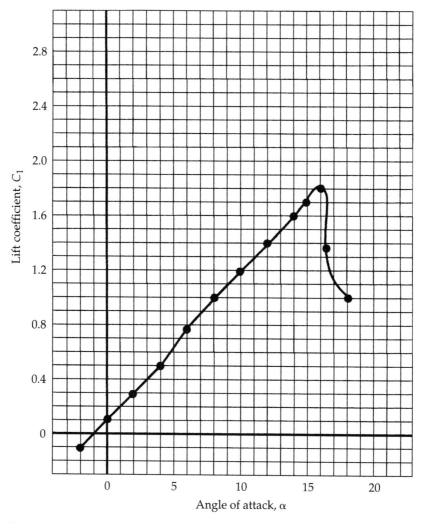

Fig. 1-13. This graph shows the lift coefficients plotted for angles of attack for the NACA 23012 airfoil.

HIGH COEFFICIENT OF LIFT DEVICES

Takeoff and landing flaps, slots, slats, vortex generators, and other devices that allow aircraft to take off and land at speeds lower than possible without these devices have been called "high-lift" devices. These devices do not, however, create lift higher than the weight of the aircraft. They do, however, allow the required lift to be created at lower speeds and so should be properly called *high-coefficient-of-lift* devices.

Flaps

The two ways that the maximum coefficient of lift of an airfoil section can be increased are by increasing its thickness or by increasing its camber. Because there is no practical way to increase the thickness of an airfoil in flight (accumulating ice will not accomplish this goal), increasing the camber of the airfoil with devices known as *camber changers* achieves the same end. These devices are commonly known as flaps and come in several different varieties (Fig. 1-14).

The plain flap is the simplest of the flaps and is simply a hinged section of the trailing edge of the wing. This flap acts solely by increasing the camber of the wing and does so well aft on the wing cord. This creates a significant increase in the maximum coefficient of lift with an equally significant increase in drag. Additionally, the zero-lift angle of attack becomes more negative.

The split flap consists of a plate deflected from the under surface of the wing and produces a slightly higher C_{LMAX} than does the plain flap. This arrangement also causes a much larger increase in drag from the great turbulent wake caused by this type of flap. This might be advantageous, however, in military aircraft—especially those based on aircraft carriers. The large increase in drag allows a much steeper approach and will require much higher power settings that will minimize engine acceleration times in the event of a balked approach.

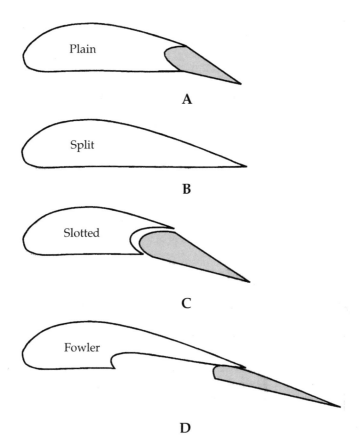

Fig. 1-14. Airfoils A through D are shown equipped with plain, split, slotted, and Fowler flaps, respectively.

Slotted flaps are similar to plain flaps with the exception of the gap between the main section of the wing and the flap that is given special contours. This shaping allows high-energy air from the lower surface of the wing to be ducted over the top surface of the flap thus providing more energy to the air flowing over the flap. This energy delays airflow separation and results in a smaller wake and less drag. The slotted flap can be extended to relatively high angles to give a good deal of lift without an excessive drag penalty.

Fowler flaps move a considerable distance aft as they are deployed, thereby increasing wing area in addition to increasing camber. Because of the aft extension, Fowler flaps require a track mechanism that complicates wing structure and adds weight. This flap is characterized by large increases in C_{LMAX} with minimum changes in drag.

The Fowler flap, which has slotted and double-slotted variations, is often used on aircraft, such as commercial airliners that require extreme increases in lift. The slotted and double-slotted Fowler flaps are high-lift producers; however, the structure is proportionately more complicated.

Flaps increase the lift coefficient of the airfoil with all other conditions being equal. Figure 1-15 shows a lift curve (C_L versus α) for a hypothetical airfoil with flaps retracted and flaps deployed. Note that for any given angle of attack, the lift

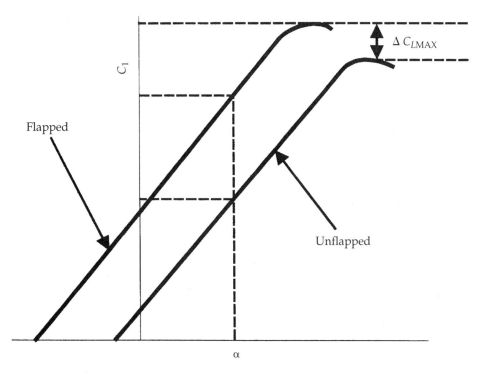

Fig. 1-15. This graph shows the lift curve for a hypothetical airfoil with and without flaps. Notice that although the lift coefficient is greater for the flapped wing at any given angle of attack, the maximum lift coefficient occurs at a lower angle of attack with flaps.

coefficient with the flaps down is greater than that with the clean wing. The significant effect of flaps is to allow the lift coefficient to increase beyond that possible without flaps. Maximum lift coefficient is reached at the stalling angle of attack regardless of flap position. Notice, however, from Fig. 1-15 that even though the stall in the flap-deployed condition occurs at a lower angle of attack, the lift coefficient is higher. This in itself is not overly significant since a higher C_L can be obtained without use of flaps by going to a higher angle of attack. A higher angle of attack, however, requires a lower velocity since lift is proportional to angle of attack multiplied by the square of velocity:

$$L \, \rho \, \alpha \times V^2$$

If one of these terms is increased for any given airplane weight and configuration, the other must decrease to produce the same result. Lowering velocity would increase angle of attack, and consequently lift coefficient, but would also eventually reach the stalling point. The higher maximum lift coefficient allows the aircraft to fly slower before it stalls. This results in slower approach speeds, slower touchdown speeds, and shorter landing distances.

At small deflection angles, flaps increase lift more than they increase drag. This can be used to advantage during takeoff when flap deflection can shorten the takeoff run. Larger angles of flap deflection create increasing amounts of drag and are not useful for takeoff because of the effect of slowing the takeoff roll. On landing, however, both the increase in lift (allowing slower approach speeds) and the increase in drag (allowing steeper approaches) are desirable and large flap angles are favored for this phase of flight.

Effectiveness of wing flaps on a wing configuration is dependent upon many different factors. The amount of wing affected by the flaps is a major consideration. Because the outboard section of the wing is usually reserved for ailerons, the actual maximum lift properties of the wing will be less than the two-dimensional section.[12] The thickness of the wing will also be a factor in determining the effectiveness of the flaps. Flaps are more effective on thicker wings. Additionally, any sweep back of the wing will cause an additional reduction in the effectiveness of the flaps.

Leading-edge devices: slats and slots

Slats and slots are the most common high-lift devices applied to the leading edge of the wing. Slots, shown in Fig. 1-16, conduct high-energy air into the boundary layer on the upper surface of the wing and delay airflow separation to some higher angle of attack and lift coefficient. Because the camber of the wing is not changed, the slot simply delays the stall to a higher angle of attack and consequently higher lift coefficient.

A disadvantage of the slot is that it creates excessive drag at the low angles of attack associated with normal cruise speed. In order to overcome this disadvantage, leading-edge segments that open at a low speed to provide the additional lift but close at higher speeds to avoid high drag are used. These leading-edge segments are referred to as *slats*. Slats add complexity and weight to the wing structure, but when properly designed, they can be made to open and close auto-

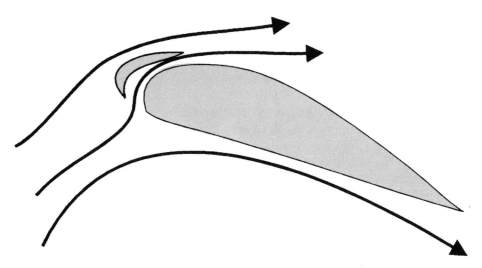

Fig. 1-16. Leading-edge slots conduct high-energy air over the upper surface of the wing at high angles of attack.

matically. In cruise configurations, the high-pressure air on the leading edge will push the slat closed. When the stagnation point is low on the airfoil at high angles of attack, the airflow will be moving upward over the leading edge of the wing as shown in Fig. 1-16. The lower pressure of this flow tends to pull the slat away from the wing leading edge opening the slot.

Vortex generators

Recall that a stall results when the boundary layer of air over an airfoil no longer possesses sufficient energy to overcome the adverse pressure gradient and separates from the surface of the wing. It follows then that if a means can be found to add kinetic energy to the boundary layer, flow separation can be suppressed, higher angles of attack can be used, and higher values of C_L can be realized. The simplest method of adding energy to the boundary layer seems to be by stirring up the boundary layer. This replaces the slow-moving laminar airflow next to the wing with faster moving (higher energy) turbulent air.

One method of producing turbulence and adding energy to the boundary layer is the use of *vortex generators*. These small airfoil-shaped vanes protrude upwards from the wing into the airstream. The tip vortices from these airfoils mix the turbulent outer layers of the boundary layer with the slow-moving laminar layers thus adding energy. Although the vortex generators increase the parasite drag, this is more than offset by the advantages of reduced takeoff and landing speeds due to the increased lift coefficient provided by the delayed separation of the boundary layer.

DRAG

Drag is the term used to denote the resistance to airflow. It is impossible to move a body through the air (or air over a body) without some drag. The process of creating lift also creates drag, and, unavoidably, the more lift there is, the more drag there is. Drag is not quite as simple as would first appear, however, and has several distinct origins and identifying terms.

Parasite drag

Pressure drag Parasite drag is the drag associated with motion of air over a surface (or the surface through the air). When you stick your hand out the window of a moving car and hold it perpendicular to the airstream, you feel the effects of pressure drag. The pressure on the upwind side of your hand is obviously greater than that in the wake of the wind and a force in the direction of the airstream results. A flat plate placed perpendicular to the airflow would experience the same pressure drag as shown in Fig. 1-17A.

The imbalance of pressure between the front face of the plate and that on the aft face results in a drag force. The larger the wake, the larger the drag. The wake can even be made larger than the bounds of the plate by curving the plate into the airstream as shown in Fig. 1-17B. If the plate is curved in the direction of the airstream, however, the size of the wake and the consequent drag will be reduced. Even curved in the direction of the airstream, however, the plate has a rather large wake and associated drag. If an aft body is attached to the plate as in Fig. 1-17D, the

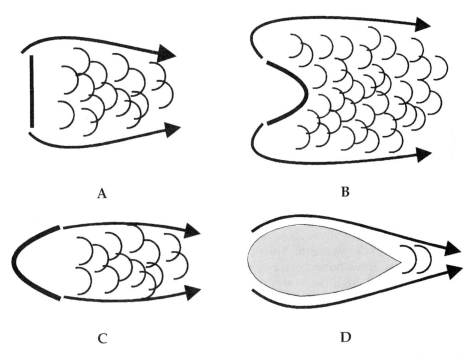

Fig. 1-17. The strength of pressure drag (or wake drag) is influenced by the shape of the body in the airstream. A body curved into the airstream, B, will experience the strongest pressure drag; a body streamlined with the airstream, D, will experience the least pressure drag.

wake can be reduced dramatically. The trailing shape of a body has a great effect on the drag, and fairing the afterbody to a gradually tapered point induces the air to flow along the trailing edge. Unfortunately, the flow will separate from the body at some point ahead of the trailing edge leaving a small but definite wake.

The reason for this airflow separation was addressed in the stalls discussion. As the airflow moves over the surface of the body, it is creating friction from the boundary layer. As we move aft along the body, we will notice that the boundary layer becomes thicker (Fig. 1-18). As the boundary layer becomes thicker, more friction drag is produced. This friction drag reduces the energy of the boundary layer. Additionally, as the flow moves aft on the body, the cross-sectional area of the body becomes smaller, and the velocity of the airstream is less; consequently, the pressure is also increasing. This is the adverse pressure gradient we talked about with stalls, and it is here working against the airflow around the aft body. Of course, the longer the airflow can be induced to remain attached to the body, the less drag will result from wake drag. The longer the flow remains attached, however, the more skin friction will be present. It should be apparent that these two types of friction are not independent because the strength of the wake drag depends upon the point of flow separation, which depends greatly on surface friction.

Friction drag One might first think that placing the plate (or hand) parallel to the airstream would greatly reduce or eliminate the drag experience. As for pressure drag, this is true, but there is still the friction drag associated with airflow. Air flowing over the surface area of any body creates friction due to the viscosity, or stickiness, of the air. All fluids have viscosity and although air is not as viscous as oil or molasses, it does have a frictional effect on a surface over which it flows. This is called *skin-friction drag* and it is but one component of parasite drag.

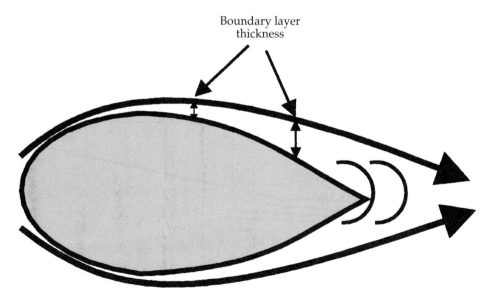

Fig. 1-18. The boundary layer thickens as the air moves aft on an airfoil.

Because of the viscosity of air, the air sticks to any surface over which it flows, causing the velocity of the air directly on the surface to be zero, regardless of the velocity of the free airstream. As we progress away from the surface of the airfoil, the velocity of the airstream gradually increases until it reaches the free airstream velocity. This distance is not large for most flight conditions—usually less than $\frac{1}{20}$ inch at the leading edge and less than $\frac{1}{2}$ inch at the trailing edge. This area between the surface and the point where the velocity reaches free airstream velocity is known as the *boundary layer*.

The smoother the air can be made to flow over the wing, the less friction drag will be created. A smooth, layered, streamlined airflow is termed *laminar flow* and is the ideal. The other type flow existing over the surface of the airfoil is turbulent flow where the streamlines break up and intermingle with each other. Figure 1-19 shows a boundary layer beginning as a laminar-flow layer and transitioning into a turbulent flow. Almost everyone has seen laminar flow transition into turbulent flow but might not have recognized the significance at the time. Picture a cigarette burning undisturbed in an ashtray in a no-wind condition. As the smoke first rises from the cigarette, it is as laminar flow. As the smoke continues to rise, however, it begins to curl and twist: turbulent flow.

The thicker turbulent layer of air will create more friction drag than the thinner, smoother laminar flow area. It is obviously a goal of aircraft designers to make the wing (and other aircraft components) as smooth and streamlined as possible in an attempt to maintain laminar flow over a surface as long as possible. This goal, while admirable, is not easily achieved. It seems that an airflow has a great tendency on its own to transition to turbulence. Any surface irregularity will enhance this tendency. Keeping the bugs off the leading edge really will make your Barnburner 126 fly faster!

Special cases of parasite drag All drag can be classified as either parasite (due to motion) drag or induced (from the production of lift) drag. Parasite drag, as we have just seen, can be further broken down into skin-friction drag and pres-

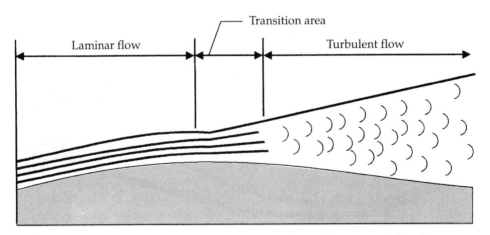

Fig. 1-19. A boundary layer begins as a laminar flow and transitions to a turbulent flow.

sure drag. There are some special cases of parasite drag, however, that do not fit neatly into either of these two divisions.

Profile drag is the wing's parasite drag. We usually think of induced drag when we speak of the wing, but it encounters skin friction and wake-pressure drag also. Profile drag will vary with angle of attack, becoming higher at higher angles of attack.

Interference drag is the drag that results from the intersection of two different bodies such as the wing and the fuselage. If the total drag of the fuselage minus the wing were measured and a similar measurement made on the wing, it would be found that the sum of the two measurements is somewhat less than the drag produced by the wing joined to the fuselage. Small vortices and the interaction of the boundary layers at the junction of the two bodies cause this increase in drag. This effect can be reduced but never entirely eliminated by filleting the junction.

Drag coefficient

As we have seen, the total parasite drag of an aircraft is the pressure, or wake, drag plus the friction drag plus interference and profile drag. Drag, like lift, is proportional to the dynamic pressure of the airstream and the area on which it acts. Also like lift, drag is not exactly equal to the dynamic pressure times the area, but is affected by the surface roughness, shape of the body, and other factors. To account for these variations, a drag coefficient, C_D, is used to describe the amount of dynamic pressure force that is converted to drag according to the following equation:

$$\text{Drag} = C_D \times q \times A$$

where: C_D = drag Coefficient
           q = $\frac{1}{2}(\rho\, V^2)$
           ρ = density
           V = velocity
           A = area

The drag coefficient is similar to the lift coefficient in that it shows how efficiently the body turns the dynamic pressure of the air into drag; however, while we are interested in achieving high numbers for the lift coefficient, the opposite is true for the drag coefficient. We want to create a design that is not at all efficient in turning dynamic pressure into drag! Notice also, that the drag will vary directly with the square of the airspeed. If the airspeed is doubled, the drag is quadrupled!

Drag coefficient can be determined by wind tunnel tests that measure the total drag of the body and solve the equation for C_D:

$$C_D = \frac{\text{Drag}}{(q \times A)}$$

Induced drag

Induced drag is a subject that is often given little time and space in ground schools and pilot manuals. Usually simply described as the "drag due to lift," it is a concept misunderstood by many aviators. Because induced drag is a direct result of wingtip vortices, it will be around as long as aircraft have wingtips, although its effects can be altered somewhat. Only a wing with infinite span would not suffer from induced drag caused by wingtip vortices.

Figure 1-10 illustrates the pressure differential between the top and bottom surfaces of the wing that induces the formation of a vortex at each tip, causing a downward push to the air leaving the trailing edge of the wing. This downward push causes the airstream to leave the wing at a slight downward angle from the incoming air, as shown in Fig. 1-20. This downward deflection of the air leaving

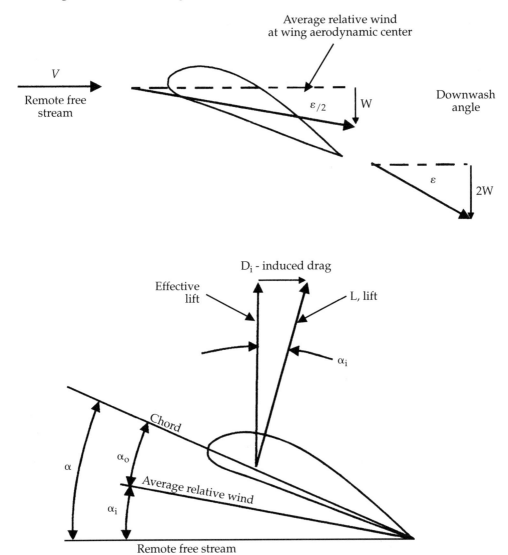

Fig. 1-20. The downwash from the wing causes a change in the average relative wind over the wing. This change has the effect of tilting the lift vector aft and increasing the angle of attack of the wing by an equal amount.

the wing also causes an aft tilt to the lift vector since it is perpendicular to the airflow. This aft tilt causes a component of the lift vector to act in a streamwise direction, opposite to the flight path. Recall that this component of the lift vector is known as induced drag.

Induced drag can be measured in the same manner as parasite drag and is described as a drag coefficient, in this case the *induced-drag coefficient*. The induced-drag coefficient, C_{Di}, is proportional to induced drag and inversely proportional to dynamic pressure and wing area as follows:

$$C_{Di} = \frac{D_i}{q \times A}$$

Because induced drag is a result of the production of lift, it is reasonable to assume that induced drag is proportional to lift, and in fact it is proportional to the square of lift as follows:

$$C_{Di} = k \times \left(\frac{C_L^2}{AR} \right)$$

The k in this equation is the constant of proportionality and varies with wing planform shape and orientation with the fuselage. The minimum value possible for k is $1/\pi$ for an elliptical wing with no fuselage present. Because an elliptical wing has the lowest drag, it follows that any other shape will have a higher value for k and hence a higher induced drag.[13]

If we continue for a moment to examine the equations that address induced drag, we can discover some interesting facts. First of all, the proportionality of induced drag to lift coefficient is significant. Consider the following equations. Lift coefficient is defined as the ratio of lift to $q \times s$ as follows:

$$C_L = \frac{L}{q \times S}$$

$$= \frac{L}{\frac{1}{2} \rho V^2 \times S}$$

The induced-drag coefficient is equal to this whole term squared and multiplied by k over the aspect ratio as follows:

$$D_i = C_{Di} \left(\frac{1}{2} \rho V^2 \right) \times S$$

$$= \frac{k}{AR} \left[\frac{L}{\left(\frac{1}{2} \rho V^2 \right) \times S} \right]^2 \times \left(\frac{1}{2} \rho V^2 \right) \times S$$

$$= \frac{k}{AR} \times \frac{L^2}{\frac{1}{2} \rho V^2 \times S}$$

Because aspect ratio, AR, can be calculated by dividing the area, S by the square of the span, b^2, and because in level, unaccelerated flight lift can be replaced by weight, we have the following:

$$D_i = k \frac{W^2}{\frac{1}{2} \rho V^2 b^2}$$

Induced drag will vary proportionally with any term in the numerator of this equation; therefore, we can see that wing planform and aircraft weight have a direct effect on induced drag. In fact, induced drag will vary with the square of the weight; doubling the weight of the aircraft will quadruple the induced drag. Also, induced drag will vary inversely with any term in the denominator of the equation. Increasing the velocity of the aircraft will decrease induced drag; in fact, doubling the velocity will cut the induced drag to one fourth the original value. Also, increasing the span of the wing will reduce the induced drag. Notice that as density increases, induced drag decreases, and conversely, as density decreases—at increasing altitude—induced drag increases. In summary, then, the factors that tend to increase induced drag are:

- High altitude
- High gross weight
- Any wing design other than elliptical (less efficient design)
- Short wingspan
- Low velocity

Those factors that occur as the square of the term such as weight, wingspan, and velocity will have a far greater effect on induced drag than will the others.

Total drag

Total drag for an aircraft is the sum of parasite drag and induced drag. Parasite drag for this purpose is considered to be all drag that is not induced. Parasite drag increases dramatically with an increase in speed. Induced drag, contrary to common sense, decreases as velocity increases. The airplane can be slowed to reduce parasite drag, but then the induced drag increases.

Total drag for an aircraft can be written as an equation as follows:

$$D_{\text{TOTAL}} = (C_{DP} \times q \times S) + (C_{DI} \times q \times S)$$

As can be seen, the two terms of the equation are the parasite drag coefficient, C_{DP}, times the dynamic pressure, q, times the wing area, S, and the induced drag coefficient, C_{DI}, times the dynamic pressure, q, times the wing area, S. (Wing area is commonly used in these equations as the representative area for parasite and induced drag for the whole airplane.) If we expand the equation using the equation that was developed in the previous section, we will have the following:

$$D = C_{D_P} \left(\frac{1}{2} \rho V^2 \right) S + k \frac{W^2}{\frac{1}{2} \rho V^2 b^2}$$

That looks like a fairly complicated equation. If we keep all terms such as weight, wingspan, area, and density constant, we can substitute constant symbols, X_1 and X_2, for the combination of those terms. The equation is now much simpler:

$$D = X_1 \times V^2 + X_2 \times \frac{1}{V^2}$$

The first term in the equation, $X_1 \times V^2$, represents parasite drag and makes it quite clear that parasite drag varies directly with the square of the velocity. The second term, $X_2 \times 1/V^2$, is induced drag and can be seen to vary inversely with the square of the velocity.

Figure 1-21 is a plot of the parasite-drag curve, the induced-drag curve, and the plot of the sum of the two curves, which represents the total drag on the aircraft. Induced drag is highest at low velocities, while parasite drag is lowest at low velocities. Notice that the point on the total-drag curve that is directly above the intersection of the induced-drag and parasite-drag curves is the lowest point on the total-drag

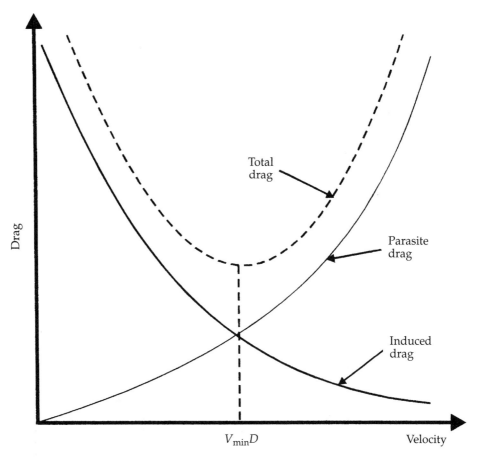

Fig. 1-21. The total drag of an aircraft is a sum of the induced drag and the parasite drag. The velocity of minimum drag, $V_{min}D$, is found at the intersection of the two drag curves.

curve. This is the speed, and consequently the angle of attack, at which the total drag on the aircraft will be the lowest. This is also the speed at which parasite and induced drag are equal. Notice that above and below this point the drag on the aircraft increases. (There will be a particular set of curves for changes in either of the terms we have considered constant. Changes in gross weight and altitude will change these curves. Hopefully the wingspan or area will not change during the course of the flight!) This point on the graph has a great deal to do with aircraft performance and will be discussed more thoroughly in chapter 8.

GROUND EFFECT

Ground effect is a term that is much bandied about in hangar-flying sessions but rarely totally understood. Many pilots will describe ground effect as a tendency to float. Some pilots and some manufacturers of low-wing aircraft insist that ground effect is a cushion of compressed air that forms between the wing and the ground. Still others will announce, very authoritatively, that ground effect is the result when air is forced downward off the wing and rebounds from the ground against the bottom of the wing.

Of all these dubious explanations, the "tendency-to-float" theory is perhaps the closest to the truth. *Ground effect* is a reduction in induced drag that occurs near the ground due to the reduction in the downwash from the wing. Obviously, any reduction in drag without a concurrent reduction in speed will cause that tendency to "float." Refer again to Fig. 1-20. Note that at low airspeeds the total drag is comprised more of induced drag than of parasite drag. Any reduction in the induced drag at this point will have a dramatic effect on the total drag.

Induced drag is caused by the vortex action resulting from the pressure differential between the top and bottom of the wing. If the wing were of infinite span, or if end plates could be installed on the wing to prevent the formation of these vortices, the vortices and consequently the induced drag could be eliminated. Neither of these alternatives is practical, however, when an aircraft flies close to the ground; the ground serves as a barrier to vortex action. Additionally, the proximity to the ground changes the downwash angle, which affects the angle of the relative airstream. Notice from Fig. 1-20 that reducing the relative airstream angle will increase the effective angle of attack. Increased angle of attack, with the airspeed constant, leads to increased lift.

Another effect of ground effect and reduced downwash is the effect on the tail of the aircraft. Downwash decreases the angle at which the airstream hits the horizontal tail. Because the tail flies at a negative angle of attack, decreasing this angle has the effect of decreasing the negative angle of attack of the tail and gives the aircraft a nosedown pitching moment and tends to increase speed and reduce drag.

As might be inferred from the name, ground effect occurs only in close proximity to the ground. In fact, the aircraft must be within one wingspan from the ground to experience this effect. At this height, the induced drag is reduced by only 1.4 percent. As the wing gets closer to the ground, the percentage increases until at ¼ span, the reduction is 23.5 percent. At ¹⁄₁₀ span, the reduction in drag is 47.6 percent.

As the aircraft descends into ground effect, the following effects will take place:

- Because of the reduced induced angle of attack and change in spanwise lift distribution, a smaller wing angle of attack will be required to produce the same lift coefficient. If a constant pitch attitude is maintained, an increase in lift coefficient will be experienced.

- The reduction in downwash due to ground effect will produce a change in longitudinal stability and trim. Generally, the reduction in downwash will increase static longitudinal stability. Additionally, additional up-elevator trim is usually required to trim the airplane at a specific lift coefficient. A nosedown pitching moment will occur. This is true for aircraft with conventional tails. T-tailed configurations are usually not affected by downwash.

- The thrust required at low speeds will be reduced as a result of the reduction in induced drag.

- Errors in airspeed and altitude indication will be noticed. In most cases, airspeed and altitude will read low due to an increase in the local pressure at the static source due to the change in upwash, downwash, and tip vortices.

WEIGHT AND CENTER OF GRAVITY

Weight and balance calculations are part of every FAA certificate's written examination. Pilots are taught to calculate the gross weight of the aircraft and the center of gravity and determine if these results are within the aircraft limitations. ATP written tests contain center of gravity shifting problems. Unfortunately, the consideration of center of gravity usually ends there for most pilots.

Limitations on aircraft weight are usually well understood (if sometimes exceeded) by most pilots. There is a general understanding that the wings can provide only so much lift and the engines only so much thrust, and therefore the weight of the aircraft must be limited within these parameters. The reasoning behind center of gravity limits is rarely considered beyond the general vague knowledge that aft locations for the center of gravity can make the airplane less stable.

Consider the wing shown in Fig. 1-22. The force of lift acts in an upward direction through the aerodynamic center of the wing. The weight of the aircraft acts in a downward direction through the center of gravity. It appears from this illustration that the wing cannot fly level and will have a nosedown pitching moment. Notice what happens when a tail is added to the configuration in Fig. 1-23. The tail of the aircraft has an aerodynamic center, but notice that the lift on the tail is directed in a downward direction. This balances the nosedown pitching moment caused by the relative locations of the center of gravity and the aerodynamic center of the wing. The lift produced by wing, times the arm (a) between the aerodynamic center and the center of gravity, is a moment equal to the moment (b) formed by negative lift produced by the tail surface times the arm between the aerodynamic center of the tail and the center of gravity.

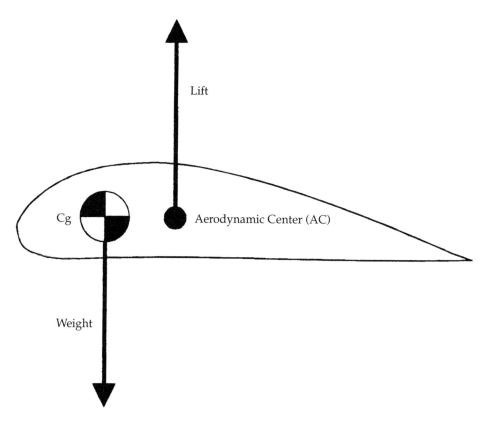

Fig. 1-22. The relative locations of the center of gravity and the aerodynamic center of a wing. Notice that this configuration induces a nosedown pitching moment.

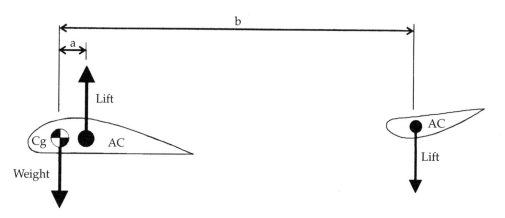

Fig. 1-23. The addition of a tail surface contributing downward (negative) lift balances the nose-down pitching moment of the wing.

As an example, consider the weight of the aircraft to be 10,000 pounds and the arm between the aerodynamic center of the wing and the center of gravity to be 10 inches. Assign the value of 343.333 inches to the arm between the center of gravity and the aerodynamic center of the tail, and assume a value of 300 pounds for the amount of lift produced by the tail. If we multiply arm b, 343.333, by the negative lift of the tail, we obtain:

$$343.333 \text{ inches} \times -300 \text{ pounds} = -102,999.9 \text{ inch-pounds}$$

The calculation for the wing results in:

$$10 \text{ inches} \times 10,000 \text{ pounds} = 100,000 \text{ inch-pounds}$$

It seems that we do not have a balanced situation until we consider the fact that the wing must produce an amount of lift to equal the weight of the aircraft *plus* the negative lift produced by the tail. If we add 300 pounds to account for the negative lift on the tail to the weight of the aircraft, we will obtain the total lift that the wing must produce to maintain level flight. The calculation for the wing moment now looks like this:

$$10 \text{ inches} \times 10,300 \text{ pounds} = 103,000 \text{ inch-pounds}$$

That's as close to equal as we will get for an illustrative example.

What will be the effect if the center of gravity is moved farther aft? Assume that the arm (a) between the aerodynamic center of the wing and the center of gravity is now only 5 inches. The moment calculation for the wing is as follows:

$$5 \text{ inches} \times 10,150 \text{ pounds} = 50,750 \text{ inch-pounds}$$

The moment calculation for the tail is:

$$338.333 \text{ inches} \times 150 \text{ pounds} = 50,750 \text{ inch-pounds}$$

As can be seen from the above calculations, moving the center of gravity aft reduces the download required of the tail to balance the nosedown pitching moment of the wing. Because the wing must produce an amount of lift equal to the gross weight of the aircraft plus the download on the tail, the aft center of gravity results in a "lighter" aircraft. Because higher gross weights require higher angles of attack and result in higher values of induced drag, loading the aircraft for the aftmost center of gravity possible within limits is the most efficient configuration. An aft center of gravity location will, as we have said, result in a reduction in the amount of negative lift required of the tail. Because production of lift always incurs the penalty of induced drag, the reduction of lift required of the tail will reduce induced drag. This hypothetical example certainly won't result in a large increase in the indicated airspeed of a corporate jet, but the savings possible with airline-sized equipment such as 747s and DC-10s are substantial.

Factors limiting center-of-gravity range

The center-of-gravity limits for most aircraft are relatively narrow in terms of percent of wing chord. Many aircraft describe the center-of-gravity limits in percentage of the mean aerodynamic chord (MAC) rather than inches. The extent of the range of allowable values of center of gravity at maximum certified weight is typically no more than 25 percent of MAC.

As the center of gravity moves forward, the aircraft becomes more stable. As the aircraft becomes increasingly more stable, it becomes less controllable and increasingly large control inputs are required to affect a change in attitude. The upper limits of stability are set by the lower limits of controllability. Recall that flying in ground effect induces a nosedown pitching moment and decreased elevator effectiveness. The forward limits of the center of gravity must be restricted to allow enough elevator effectiveness to rotate the aircraft to a flying angle of attack while in ground effect and to rotate the aircraft to a landing attitude in ground effect. Because the tail must provide negative lift to balance the nosedown pitching moment of the wing, this fact will also limit the location of the forward center of gravity limit. The tail can provide only a finite amount of lift.

Because of the limits of the tail surfaces, the forward limits of the center-of-gravity envelope for most aircraft are restricted at higher gross weights. The arm through which the weight works must be restricted to ensure that the tail is capable of producing the necessary balancing force. Figure 1-24 shows the center-of-gravity envelope for the Learjet model 36A. Notice that the center-of-gravity limits move forward as weight is reduced.

As the aircraft flies out of ground effect, the nosedown pitching moment will be relieved and the elevator at that point must be able to counter the resultant pitchup. As the center of gravity moves farther aft, the aircraft will become less stable, but more controllable. The lower limits of stability may be set by the upper

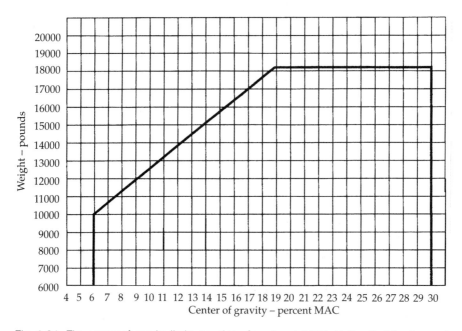

Fig. 1-24. The center-of-gravity-limits envelope for a Learjet 36A. Notice that the forward limit of the center of gravity slopes aft at increasing gross weights.

34 *A review of basic aerodynamics*

limits of controllability. If the center of gravity is allowed to move aft of the center of pressure, the role of the tail will be reduced and the tail surfaces will now be required to produce lift in the upward direction. This leads to an extremely unstable condition.

Endnotes

1. This is the airfoil found on the outboard section of the wing of a Cessna Citation CE-500 at wing station 247.95.

2. Airfoils of this series were first used on the P-51 Mustang for their low-drag characteristics.

3. Actually, the airfoil does produce lift, but the "lift" produced by the curved top surface of the airfoil is exactly balanced and canceled by the "lift" generated by the bottom-curved surface.

4. Bernoulli actually addressed the behavior of fluid streams through a ducting. More will be said about his studies in chapter 4.

5. It is, in fact, known as the kinetic energy of the air, or the energy of motion.

6. *Momentum* is a physical quantity defined as mass times velocity. An increase in mass with velocity held constant will increase momentum. Also, an increase in velocity with mass held constant will increase momentum.

7. As anyone who has been in aviation for more than two days knows, the thing that makes airplanes fly is not air pressure differentials anyhow. The thing that makes airplanes fly is cubic money.

8. Any vector can be projected along the axes of a rectangular coordinate system and completely described by its components. The projection of a vector, V, along the x (horizontal) axis of the coordinate system is called V_x and its projection along the y (vertical) axis is termed V_y. The angle the vector makes with the x axis is termed θ. The components of the vector are then given by:

$$V_x = V \cos \theta$$
$$V_y = V \sin \theta$$

See any beginning physics or trigonometry text for a more thorough treatment of vectors.

9. Do not confuse angle of attack with angle of climb or deck angle. The angle of attack is the angle between the chord line of the wing and the relative wind. The angle of climb (or descent) is the angle of the flight path with the ground. The deck angle is the angle of the longitudinal axis of the aircraft with the horizon. The angle of attack can be high with the deck angle below the horizon. I know this was taught in your first flight training, but it bares repeating.

10. This is the type of stall-warning device used on the Cessna Citation 500 and 550 aircraft. The stall strip causes a portion of the wing in front of the horizontal stabilizer to stall. The turbulent airflow causes the horizontal stabilizer to buffet warning of an impending stall: low tech but effective.

11. This graph approximates the graph for the NACA 23012 wing section. Data on various airfoil sections and the theory of wing sections can be found in the

book *Theory of Wing Sections Including a Summary of Airfoil Data* by Abbott and Doenhoff.

12. Some STOL kits installed on various aircraft use aileron "droop" to increase the area of the flaps. When flaps are deployed, the angle of the aileron is changed to somewhat compensate for the less-than-full-span flap.

13. The British Spitfire used in World War II was a notable design with an elliptical wing. Labor costs and complexity of construction have made elliptical wings impractical and outweighed their advantage. Tapered wings are almost as efficient as elliptical wings and are far less complex and costly to construct.

2

High-speed and high-altitude aerodynamics

AT SLOW FLIGHT SPEEDS, THE STUDY OF AERODYNAMICS IS GREATLY simplified by the fact that air can be considered to be incompressible; the air can undergo changes in pressure without apparent changes in density. Airflow in these low speed ranges is similar to the flow of water, hydraulic fluid, or any other incompressible fluid. During high-speed flight, however, the pressure changes that take place are much larger than those addressed previously and significant changes in air density occur: compressibility.

THE SPEED OF SOUND

It is hard to imagine anyone alive in the modern world who has not heard about the speed of sound and the "sound barrier," yet these terms, like aviation itself, are less than a century old. It had been observed shortly before World War II that as airplanes reached speeds of 300 to 400 miles per hour, significant density changes occurred in the airstream surrounding the aircraft and drag dramatically increased.

Many leading aeronautical scientists theorized that as airplanes approached the speed of sound, the drag would increase infinitely, thus preventing further acceleration regardless of the thrust available; therefore, the term "sound barrier" came into being. Gen. Chuck Yeager was the first aviator to fly faster than the speed of sound and so was the first to break this mythical barrier.[1]

You might have wondered what sound has to do with the drag experienced by an airplane in flight. Actually, sound itself has nothing to do with it. The significance is that sound is the result of pressure disturbances in the air. For example, as a drum is struck, the vibrating drum head collides with molecules of the atmosphere and sets up pressure disturbances in waves flowing out from the source of the vibration, much like the waves of disturbance on the surface of water when a stone is thrown in. These pressure disturbances, which are transmitted by collisions between molecules in the atmosphere, finally reach our eardrums. Our eardrums vibrate and this vibration is perceived as sound.[2] The speed of sound is the rate at which small pressure disturbances will be propagated through the atmosphere. This rate of this propagation is solely a function of temperature of the air.

Temperature is a measure of the molecular energy of a substance. Air with a higher temperature has a higher molecular energy, and its molecules are moving at a relatively high rate of speed. The speed of movement causes far more collisions between molecules and allows any pressure disturbances in the atmosphere to be more easily transmitted. As the air temperature drops, the rate of collision between the molecules drops. The speed of sound, being based on the propagation of pressure disturbances in the air, is dependent upon air temperature. Table 2-1 shows the speed of sound at varying altitudes and temperatures in the standard atmosphere. The speed of sound will vary with temperature according to the following formula:

$$V_{SOUND} = V_{SOUND_{SL}} \sqrt{\frac{T}{T_{SL}}}$$

Where V_{SOUND} is the speed of sound to be found, $V_{SOUND\ SL}$, is the speed of sound at sea-level standard temperature, T is the temperature at the altitude where the speed of sound is being calculated, and T_{SL} is the standard temperature at sea level.

Table 2-1. Speed of sound at varying altitudes in standard atmosphere

Altitude feet	Temperature °C	Speed of sound (knots)
Sea level	15.0	661.7
5,000	5.1	650.3
10,000	−4.8	638.6
15,000	−14.7	626.7
20,000	−24.6	614.6
25,000	−34.5	602.2
30,000	−44.4	589.6
35,000	−54.3	576.6
40,000	−56.5	573.8
50,000	−56.5	573.8
60,000	−56.5	573.8

As an aircraft moves through the airmass, velocity and pressure changes occur that create disturbances in the surrounding airflow. These pressure disturbances are propagated through the air at the speed of sound. If the aircraft is moving at a speed slower than the speed of sound, the pressure disturbances move ahead of the aircraft, and the airflow immediately in front of the aircraft is influenced. This pressure "warning" can be seen in the typical subsonic flow patterns about an airfoil. A change in airflow direction and an upwash occurs well ahead of the airfoil.

If the aircraft is traveling at some speed faster than the speed of sound, the airflow ahead of the aircraft will not be influenced by the pressure field on the aircraft because the pressure disturbances cannot be propagated ahead of the craft. As flight nears the speed of sound, a compression wave will form at the leading edge and changes in velocity and pressure will take place quite abruptly. Typical subsonic and supersonic flow patterns are shown in Fig. 2-1.

Typical subsonic flow pattern

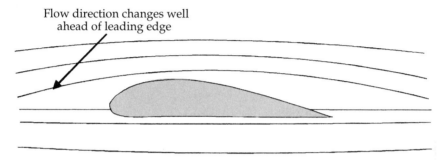

Flow direction changes well
ahead of leading edge

Typical supersonic flow pattern

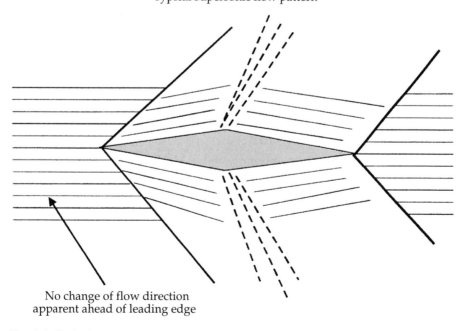

No change of flow direction
apparent ahead of leading edge

Fig. 2-1. Typical subsonic and supersonic flow patterns.

The compression wave formed on the leading edge of an aircraft flying near or faster than the speed of sound is much like the bow wave formed by a high-speed boat. If the boat moves slower than the propagation speed of the waves, no bow wave is formed. As the ship nears the speed of wave propagation, a bow wave will begin to form that will become stronger with increasing speed.

MACH NUMBER

It should by now be apparent that all compressibility effects depend upon the relationship of airspeed to the speed of sound. The term used to describe this relationship is the *Mach number*, named after the Austrian physicist Ernst Mach. Mach number, M, is the ratio of the true airspeed of an object to the speed of sound expressed as a decimal number.

$$M = V/a$$
where: M = Mach number
V = true airspeed, knots
a = speed of sound

An aircraft traveling at half the speed of sound is flying at M 0.5; a speed of M 2.0 would be twice the speed of sound. This Mach number is true only for the altitude (and consequently temperature) at which the aircraft is flying. The true airspeed yielding Mach 0.5 at sea level would translate to M 0.57 at 35,000 feet at standard temperatures.

Flying at exactly Mach 1 is a tricky proposition. Only very slight disturbances will cause the speed to go subsonic or supersonic. Also, even if the aircraft is able to maintain flight at exactly Mach 1, airflow over parts of the aircraft will exceed the speed of sound while that over other parts will be definitely subsonic. For this reason, we usually classify regimes of flight approximately as follows:

- Subsonic: Less than M 0.8
- Transonic: Mach 0.8 to M 1.2
- Supersonic: Mach 1.2 to 5.0
- Hypersonic: Greater than M 5.0

Some authors will list the transonic range as M 0.7 through M 1.3; however, it is generally agreed that the speeds 20 to 30 percent on either side of M 1.0 will be included in the transonic range. The major characteristic of flight in this regime is that airflow over different parts of the aircraft might be subsonic, sonic, or supersonic. Also, airflow in the hypersonic region has distinct characteristics from those at lower Mach numbers.

The principal differences between subsonic and supersonic flow are due to the compressibility of the supersonic flow. Any change in velocity or pressure of a supersonic flow will produce a related change in density of the air. Figure 2-2 provides a comparison between subsonic incompressible flow and supersonic compressible flow. It is assumed that continuity of flow exists through both tubes and that the mass flow at any point along either tube is constant.

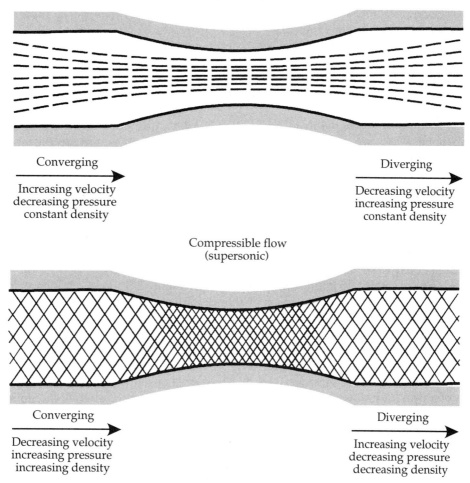

Incompressible flow
(subsonic)

Converging

Increasing velocity
decreasing pressure
constant density

Diverging

Decreasing velocity
increasing pressure
constant density

Compressible flow
(supersonic)

Converging

Decreasing velocity
increasing pressure
increasing density

Diverging

Increasing velocity
decreasing pressure
decreasing density

Fig. 2-2. Subsonic and supersonic flows through a converging and diverging tube. Note that there is a change in density with the supersonic flow. This is necessary to keep the same mass flow at a decreasing velocity.

Subsonic, incompressible flow is characterized by constant density throughout the tube. As the flow approaches a constriction (convergence), the flow expands (pressure decreases), and the velocity increases. As the subsonic flow enters a diverging section of the tubing, the velocity decreases and the static pressure increases, but the density remains unchanged.

The supersonic flow behaves differently. Changes in density are related to changes in velocity and pressure. As a supersonic flow approaches a constriction in the tube, the velocity decreases while the static pressure increases. The density of the flow must increase to keep the mass flow past any point in the tube constant.

As the supersonic flow enters the divergence in the tube, the velocity increases, and the pressure and density decrease.

Compressibility effects are ignored below approximately Mach 0.3. These effects become increasingly important as speed increases. Compressibility effects do not suddenly arrive at the critical Mach number for the aircraft, they slowly build with increasingly noticeable effects until the transonic speed range is reached at which point the compressibility effects become a major factor to be reckoned with. In general, compressibility is an aid to lift until the airstream over the wing exceeds the speed of sound.

To understand the nature of compressibility, we must return to our discussion of sound waves and waves in general. When an object moves through the air, it continuously creates small pressure waves or small disturbances in the air that spread out in all directions from the object. These expanding pressure waves travel away from the object at a constant rate of speed: the speed of sound. Although these pressure waves travel in all directions, the direction we will consider is the direction in which the object is traveling.

As long as the object, or aircraft, is traveling at a speed below about Mach 0.3, the pressure wave will travel well ahead of the aircraft, and compressibility effects will not be noticed. The pressure wave causes the air particles to move out of the way and smoothly streamline around the aircraft. Figure 2-3 shows what happens as the aircraft approaches the speed of sound. As the speed increases, the aircraft comes closer and closer to the air particles before they are "warned" by the advancing pressure wave and can change direction. The greater the aircraft speed, the fewer air particles will be able to move out of its path. As a consequence, the air particles begin to pile up in front of the aircraft, and the air density increases.

When the speed of the aircraft reaches the speed of sound, the pressure waves cannot warn any air particles in front of the aircraft because the aircraft is keeping up with the pressure wave. The aircraft collides with the air particles piled up in front of it before they can move out of the way. As a result of these collisions, the speed of the airstream directly in front of the aircraft slows down rapidly, while the density and pressure increase accordingly.

As the speed increases beyond the speed of sound, the pressure and density of the air directly in front of it are increased correspondingly. This region of compressed air extends for a distance ahead of the aircraft; the exact distance is dependent upon the size of the aircraft, the aircraft speed, and the temperature of the air. Because of the sudden nature of the changes in temperature, velocity, density, and pressure that the airflow is compelled to undergo, the boundary line between the undisturbed air and the region of compressed air is called a *shock wave*.

Critical Mach number

Recall from the discussion of lift that a gradual decrease in pressure occurs over the top surface of the wing from the leading edge to the point of maximum thickness. This decrease in pressure is due to an increase in velocity of the airstream over the airfoil. The local velocity in any part of the airstream over this part of the airfoil will be greater than that of the free airstream and will, by definition, be at a

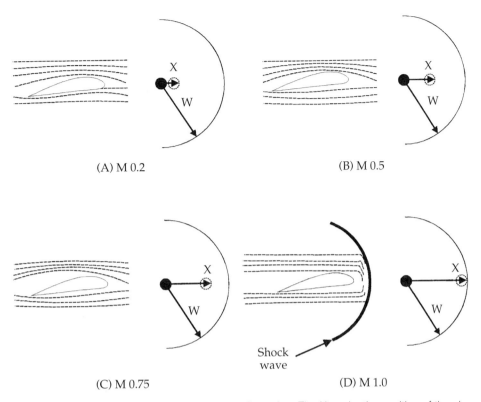

(A) M 0.2

(B) M 0.5

(C) M 0.75

(D) M 1.0

Shock wave

Fig. 2-3. Airflow distribution and pressure-wave formation. The X marks the position of the air-craft moving at the specified Mach number. The W indicates the position of the pressure wave at the time the aircraft is at position X. In D, the aircraft is moving at M 1.0, the position of the aircraft has caught up with the pressure wave and a shock wave has formed in front of the aircraft. No-tice that the airflow around the airfoil is increasingly disturbed as the speed approaches M 1.0.

higher Mach number. It follows, then, that the velocity of the airstream over the wing will reach Mach 1 before the free airstream velocity (actually, the velocity of the aircraft) reaches this point.

The free airstream velocity that will cause the airstream over the wing to reach M 1.0 is termed the *critical Mach number*. Figure 2-4 illustrates an airfoil moving be-low and at its critical Mach number. When the velocity of the free airstream is be-low the critical Mach number, in this case M 0.84, airflow over all parts of the airfoil is subsonic. When the velocity increases to M 0.84, however, the velocity over the thickest part of the wing just reaches Mach 1.0.

Critical Mach number, M_{CRIT}, is an aircraft limitation to be heeded. If the air-craft is flown at speeds in excess of its critical Mach number, numerous unsettling and potentially disastrous events occur. Before we explore these phenomena, how-ever, let us look into the formation of shock waves on airfoils.[3]

Once the aircraft exceeds M_{CRIT}, it is flying in the transonic range. Supersonic airflow exists over the wing at the area of maximum thickness. Subsonic flow ex-ists elsewhere. The pressure disturbances behind the sonic flow cannot be propa-

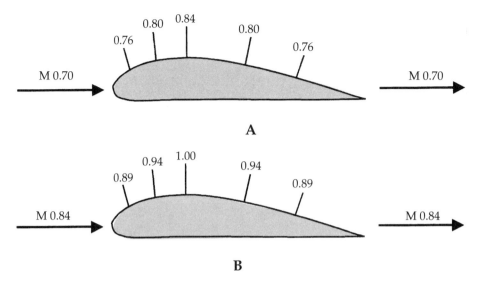

Fig. 2-4. Airfoil A is flying below its critical Mach number. As the air changes velocity over the airfoil, it does not reach M 1.0. Airfoil B is flying at its critical Mach number; airflow over the thickest part of the airfoil just reaches M 1.0.

gated forward because they run into sonic velocities traveling rearward. A normal shock wave, shown in Fig. 2-5, will form on the wing where the air slows from supersonic to subsonic.

As the air flows through a normal shock wave, it undergoes rapid compression. The compression decreases the dynamic pressure of the airstream and converts it into a static pressure rise and a temperature rise behind the shock wave. This heat rise is either radiated into the atmosphere or absorbed by the wing.[4] In either case, it is a loss of energy that must be compensated for by the engines.

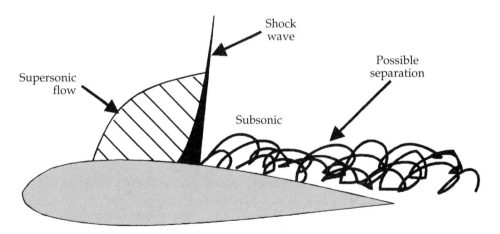

Fig. 2-5. A normal shock wave is formed as an airfoil exceeds its critical Mach number.

As we discussed in chapter 1, as the air flows aft on the airfoil, it encounters an adverse pressure gradient—an area where the pressure is increasing rather than decreasing. When this adverse pressure gradient becomes steep enough, it causes airflow separation from the airfoil. This pressure gradient is gradual in low-speed flight; however, the formation of a shock wave causes an abrupt pressure rise and a separation of the airflow from the wing. This induces a separation of the boundary layer and a wake behind the airfoil. The combined effect of the energy losses through energy dissipation and wake formation is called *wave drag*.

At speeds just slightly above M_{CRIT}, wave drag is due mostly to energy dissipation and is rather small. As the speed increases, however, the drag increase becomes significant. At a speed approximately 5 percent above M_{CRIT}, the normal shock wave on the top of the wing causes the boundary layer to separate from the wing. This causes a change in the aerodynamic force coefficients C_L and C_D, and the drag begins to rise dramatically. This number is known as *drag-rise Mach number* or *critical-drag Mach number*, M_{CRD}.

At the same time that the drag rises dramatically, the lift decreases dramatically. Shock-induced airflow separation causes a local stall situation to exist aft of the pressure wave. This causes the center of pressure of the wing to shift forward on the wing causing a pitchup. Figure 2-6 shows a graph of the effects on the coefficients of drag and lift caused by flight beyond M_{CRD}.

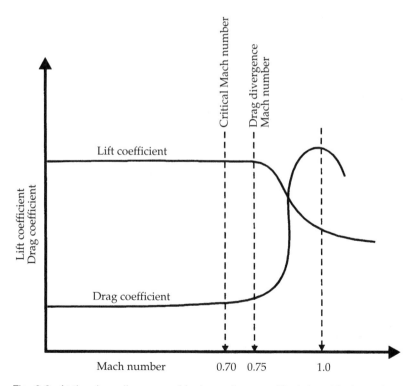

Fig. 2-6. At the drag-divergence Mach number, or critical-drag Mach number, drag increases dramatically, and lift decreases.

It is no wonder that the scientists of the day believed that drag would rise infinitely at the speed of sound. Not until flight had been accomplished faster than the speed of sound was it learned that the drag decreases dramatically after Mach 1 is achieved. A tremendous amount of thrust is required to even approach Mach 1, and propeller driven aircraft were not capable of the job. Not only are propellers airfoils that suffer the same high-speed problems as wings, but the propeller has the added complication of speed from its rotation, which causes it to suffer from shock waves and their associated drag at fairly low forward speeds. It was not until the development of the jet engine that speed in excess of sound was even remotely possible. The jet engine provides relatively high amounts of thrust at low weights and does not incur the drag associated with propellers.

As can be seen from Fig. 2-6, flight in the transonic region is not very practical. Because of the dramatic drag rises in this area, flight either below critical Mach or well above the speed of sound are the only feasible alternatives. Supersonic flight has not been deemed economical, with the sole exception of the Concorde, and commercial air travel has been confined to the subsonic realms. Transonic flight is the subject of much current research, but flight in this region is tricky at best.

TRANSONIC FLIGHT

Problems encountered in the transonic region foiled many early attempts to "break the sound barrier." Many of these problems and their associated hangar lore have been the source of the still persistent rumors that jets are hard to fly. When these problems were more thoroughly understood, several important design changes were made, and today's supersonic aircraft have little difficulty passing through this speed region on their way to the region beyond the sound barrier.

These design changes are not incorporated into aircraft designed for subsonic flight, however, and transonic flight in these aircraft will encounter all the difficulties first faced by those early jet pioneers. Flight beyond critical Mach number, whether intentional[5] or inadvertent due to turbulence or maneuvering flight, will provide many an interesting moment for the uninitiated.

As the normal shock wave first appears on the wing, the air passes through the shock wave in the direction of the airstream for a short distance. As the aircraft's speed continues to increase, the shock wave begins to move toward the trailing edge of the wing and separation of the airstream begins to occur immediately behind the shock wave. The air now begins to tumble in a random fashion, much as it does during a stalled condition.

When this turbulence begins to grow, it makes itself known to the pilot by several disconcerting and potentially lethal compressibility effects. This turbulence begins to occur when the limiting Mach number is reached and serves as a distinct warning that unless the pilot takes immediate action to reduce speed, there is much worse to follow. Some of the earlier and more common compressibility effects are:

Unaccountable roll

The separation of the airflow behind the shock wave causes the center of pressure of the wing to be shifted forward and causes a noseup pitching moment. If one wing develops a pitching moment before the other due to small differences in structure between the wings (more dents on one side, or a nonflush inspection panel, or control surfaces, etc.), a rolling moment can be induced toward the wing with the earlier development of the shock wave. If a wing drops, the aircraft will tend to yaw in that direction, and a condition similar to Dutch roll might be encountered. Also, any tendency to roll that had previously been corrected with aileron trim will recur when the trim tabs become ineffective in the turbulent airflow on the trailing edge of the wing.

Aileron buzz

Boundary layer separation on the control surfaces might cause the surfaces to rapidly oscillate, which is called *buzz*. This can cause metal fatigue problems in both the control surfaces and the hinge fittings joining the control surface to the wing.

Control effectiveness

Shock-induced separation of the airflow over control surfaces causes them to be in an area of turbulent, nonstreamlined air that causes a loss of effectiveness. Also, control surfaces act by changing the airflow around the whole wing or tail surface in front of the control surface. When there is a shock wave in front of the control surface, deflection of the surface cannot influence the airflow in front of the shock wave; therefore, any minimal aerodynamic force developed by the control surface is limited to the control surface itself and the area behind the shock wave. The separation between the horizontal stabilizer and the elevator is known to induce shock wave formation at this point, effectively blanking the elevator.

Wing twist

The wings might begin to twist due to compressibility effects. The airfoil shape over the length of the wing is seldom constant, and the differing onset of formation of shock waves and movements of centers of pressure cause this effect. This effect is unnerving at the least. (If you don't notice it, I guarantee a passenger will.)

These are general symptoms of compressibility effects. They will obviously vary with aircraft design, but they are warnings to the pilot that speed must be reduced below the limiting Mach number. The pilot might reduce power, reduce the rate of descent, or possibly use speed brakes to bring the speed back below the limiting Mach number. A word of caution here, however. It is imperative (as in all flying) that the particular characteristics and limitations of the speed brakes installed be known and thoroughly understood. Some types, mainly those that act as spoilers on the upper wing surface, might cause heavier buffeting; others might cause considerable change in pitch attitude and angle of attack.

The two most common corporate jets are totally opposite in this respect. For the Cessna Citation 500 series of aircraft, there is no limitation on deployment of speed brakes, and the appropriate action in the event of an overspeed (unlikely though that might seem to some) condition would be extension of the speed brakes. The Learjet 36 flight manual, however, contains the warning that deployment of the spoilers at speeds in excess of V_{MO}/M_{MO} is prohibited due to the significant nosedown pitching moment associated with spoiler deployment.

Should the pilot choose to ignore these early, and relatively mild in comparison, warnings and compressibility effects, more serious effects will appear. The separation of the airflow over the wings will begin to produce severe turbulence, and this will cause even more trouble.

Buffeting

Severe buffeting will almost surely be the next compressibility effect that is sure to show up. The extremely turbulent airflow separated from the wings begins to bang against the tail surfaces in a manner that is violent and irregular and disturbs the flow patterns around these surfaces causing them to buffet. This will not only be felt in the aircraft as a whole, but the control column and rudder pedals will buffet violently. This buffeting, if allowed to continue, has been known to cause separation of the tail from the aircraft.

Mach tuck

If all this has not caused the pilot to slow down and escape compressibility effects by now, a more violent effect is about to occur. Loss of longitudinal stability will cause the aircraft to "tuck under." Several factors are at work that cause this tuck under or Mach tuck:

- The downwash on the horizontal stabilizer will be decreased when the airflow over the wing separates. This causes a loss of the normal stabilizing force of the tail surfaces, and a strong nosedown pitching moment results.
- The movement of the shock waves toward the rear of the wing causes the airflow separation point to move rearward as does the center of pressure of the wing. This causes a nosedown pitching moment.
- The aerodynamic center of the wing moves aft from the quarter chord position to the 50-percent chord point as the aircraft approaches supersonic flight. All aircraft flying supersonically experience a nosedown pitching moment.

The important point is that the first signs of compressibility effects call for immediate pilot action. The airspeed must be reduced, and the nose must be eased up. The power reduction has to be fast, for when the tuck starts and the aircraft starts into a dive, the situation is going to get rapidly worse. The increased speed will cause the separation of the airflow to become more pronounced, and the severity of the buffet will become greater. The greater the turbulence over the tail, the greater will be the elevator angle and the stick force required to pull out of the dive. Some have not been successful in this maneuver, and some have lost the tail of the aircraft before they had time to begin recovery.[6]

Figure 2-7 is a graph of the stick forces required to maintain a subsonic aircraft in level flight with increasing Mach number. Notice that this is in level flight, not the recovery from a Mach tuck. As the speed increases, a pushing force is initially

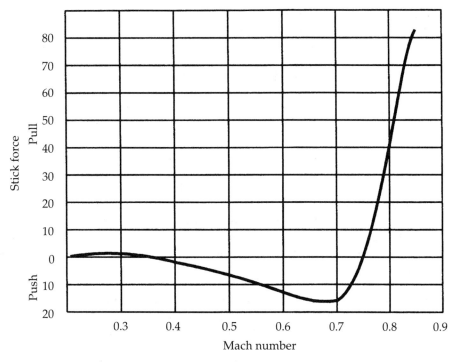

Fig. 2-7. The stick forces required to maintain level flight as Mach number increases at first require forward pressure on the yoke, but soon after the critical Mach number, the forces reverse, and rapidly increasing back pressure is needed.

needed to maintain level flight. This is fairly normal subsonic behavior. This pushing force reaches a maximum at about M 0.7, at which point compressibility effects begin to appear and the force required to trim the aircraft reverses and increases in magnitude rapidly. The dramatic and rapid increase of the force required illustrates the necessity for prompt corrective action when compressibility effects first make themselves known.

So far this discussion has addressed the effects of exceeding limiting Mach number in level flight. These effects will be more pronounced and violent, however, if the limiting Mach is exceeded in maneuvering flight, especially at high altitudes. Consider the following. An aircraft, flying at M_{CRIT} near its maximum authorized altitude, is proceeding along J-110 toward the Clovis (CZQ) VOR at which point the flight will proceed northwest bound on J-65. The inbound course to CZQ is 266°. J-65 proceeds outbound on the CZQ 304° radial. If the pilot is somewhat ham-handed and abruptly rolls into a 30° bank while maintaining altitude, the result will likely be a high-speed Mach buffet.[7]

The increased angle of attack required to maintain level flight in a turn will cause the airflow over the wing to accelerate and will produce separation effects. Additionally, any Mach tuck will proceed more rapidly due to the ease of rapid acceleration in the thin air at high altitude. This does not make a jet harder to fly, however. It just points out that limitations must be understood and obeyed. No aircraft is likely to be tolerant of rough handling near the edges of its performance envelope. Jet aircraft simply have more dramatic means of announcing their displeasure.

Maneuvering loads can also be imposed on the aircraft as a result of an unanticipated encounter with clear-air turbulence. Although the current state of technology in weather reporting and forecasting has made an unanticipated encounter with CAT much more unlikely than in the early days of high-altitude jet flight, the possibility still exists. There were many incidents in the early days of commercial jet flight of "jet upset" that were not well understood. Subsequent research and design improvements have taken the mystery out of this phenomenon.

This discussion of high-speed characteristics has been generalized. Of course, before flying any aircraft, a wise pilot will be thoroughly familiar with the particular handling characteristics of that make and model.[8] I'm sure you have decided by now that tuck-under is a dangerous situation to be avoided at all costs. Some aircraft might display tuck-up characteristics also. This can appear as a secondary characteristic after a tuck-under. As the aircraft is in a fast, steep dive with the speed increasing, a point is reached at which the turbulent airflow over the tail suddenly produces tuck-up. This will produce a violent pull out from the dive that the structure might well not be designed to withstand.

If all of this discussion has caused you concern, please let me set your mind at ease. Only the uninformed will be likely to get into any trouble over compressibility effects. The obvious solution is to observe the aircraft limitations. Aircraft designers have given much thought to the undesirable compressibility effects and there are design features incorporated into jets that are not found in slower craft.

DESIGN FEATURES FOR HIGH SUBSONIC FLIGHT

Each new generation of jet aircraft rolling off the assembly line seems to arrive with a higher critical Mach number than the previous models. Aircraft designers are continually making planes both faster and safer by incorporating design features that raise the limiting Mach numbers and lessen compressibility effects. These design features can easily be seen by observing modern jet aircraft.

Sweepback

One of the most obvious and most common ways of increasing M_{CRIT} is to sweep the wings back. Sweepback produces an unusual effect on the high-speed characteristics of a surface and has a basis in a very fundamental concept of aerodynamics. Figure 2-8 illustrates a swept wing with the free- airstream velocity broken into the components perpendicular to and parallel to the leading edge of the wing. The component perpendicular to the leading edge is less than the free-stream velocity (by the cosine of the sweep angle) and is the velocity component that determines the magnitude of the pressure distribution and consequently the critical Mach number.

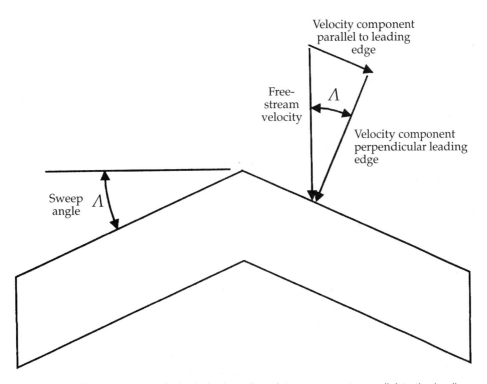

Fig. 2-8. The free-airstream velocity is broken down into components parallel to the leading edge of the wing and perpendicular to the leading edge. Because of the sweepback angle, the component of free-stream velocity perpendicular to the wing is less than the total free-stream velocity.

The component of speed parallel to the leading edge moves across airfoil sections of constant thickness and so does not contribute to the pressure distribution on a swept wing. Because only the component of speed perpendicular to the leading edge contributes to pressure distribution on the wing, higher flight speeds can be attained before critical conditions are produced on the wing. There will be an increase in critical Mach number, force-divergence Mach number, and the Mach number at which drag will increase. For example, if the M_{CRIT} of the airfoil is M 0.75 and the sweep back angle is 30°, an aircraft speed of M 0.86 might be attained before this limitation is reached. This is calculated according to the following:

$$M_{CRIT} = COS \wedge M$$

$$M = \frac{M_{CRIT}}{COS \wedge}$$

Substituting the appropriate values gives:

$$M = \frac{0.75}{COS \, 30°}$$

Design features for high subsonic flight 51

$$M = \frac{0.75}{0.866}$$

$$M = 0.866$$

Sweep will not only delay the onset of compressibility effects but will reduce the magnitude of the changes in lift and drag coefficients due to compressibility. Because the component of the free-stream velocity perpendicular to the leading edge of the airfoil is less than the free-stream velocity, the magnitude of all pressure forces on the wing will be reduced by approximately the square of the cosine of the sweep angle. Because compressibility and force divergence are a result of pressure distribution, the use of sweepback will dampen these effects. In general, sweepback delays the drag rise and reduces the magnitude of the drag rise. It will also have the effect of delaying the decrease in lift coefficient and reducing its magnitude. The graphs in Fig. 2-9 illustrate these two effects.

The use of sweepback on a transonic aircraft will reduce and delay drag rise and preserve the maneuverability of the aircraft in transonic flight. Small amounts of sweepback produce very little benefit. Sweep angles of 30–35° must be used to produce any significant benefit. Table 2-2 shows the increasing benefits to be realized with increasing sweepback.

There are also some less significant advantages of sweepback. The wing-lift curve slope is reduced for a given aspect ratio. Any reduction in the lift curve slope implies that the aircraft will be less sensitive to changes in the angle of attack. This effect is beneficial only when the effect of gusts and turbulence is considered. The

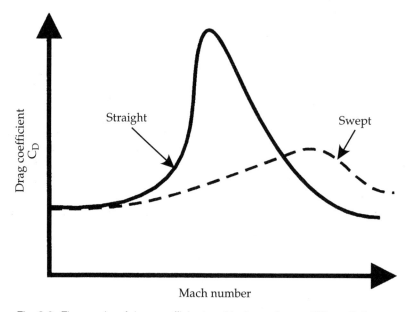

Fig. 2-9. The graphs of drag coefficient vs. Mach number and lift coefficient vs. Mach number are plotted for both straight wing and swept wing. Sweepback delays the onset of drag rise and loss of lift, and it reduces their magnitude.

52   *High-speed and high-altitude aerodynamics*

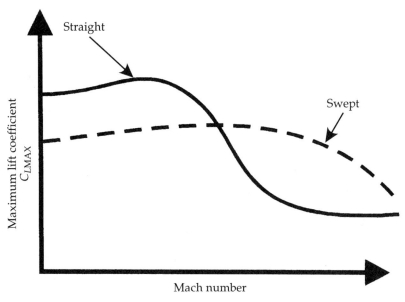

Fig. 2-9. Continued.

Table 2-2. Benefits of sweepback for varying angles

Sweep angle ($\wedge$)	Percent increase in critical Mach number	Percent increase in drag peak Mach number	Percent reduction in drag rise	Percent reduction in loss of C_{LMAX}
0°	0	0	0	0
15°	2	4	5	3
30°	8	15	15	13
45°	20	41	35	30
60°	41	100	60	50

swept wing will experience less bump due to gust for a given aspect ratio and wing loading and will give a smoother ride.9

Sweepback tends to stabilize the surface of the wing by "trailing" and raises the divergence speed. Divergence of a surface is an aeroelastic problem that can occur at high dynamic pressures. Combined bending and twisting deflections interact with aerodyvamic forces and can produce sudden failure of the surface at high speeds.

Static directional stability or weather cocking stability of an aircraft is increased slightly by sweepback. Figure 2-10 shows a swept wing in a sideslip or yaw. The wing into the wind has less sweep and therefore incurs a slight increase in drag. The wing away from the wind has more sweep and less drag. The combined effect of

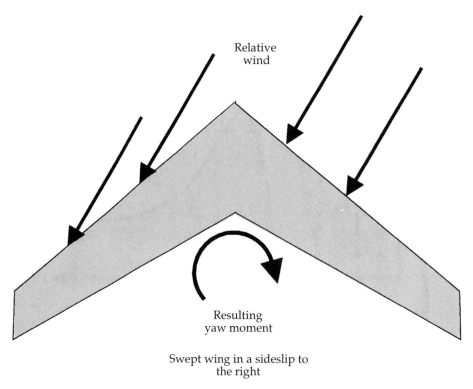

Relative
wind

Resulting
yaw moment

Swept wing in a sideslip to
the right

Fig. 2-10. A swept wing in a sideslip tends to yaw and turn the nose toward the relative wind (weathercock).

these force-changes is production of a yawing moment tending to return the nose to the relative wind. This contribution is small and is mainly of value to tailless aircraft.

Sweepback contributes to lateral stability in the same sense as dihedral. When a swept-wing aircraft is placed in a sideslip, the wing into the wind experiences an increase in lift since the sweep is less and the wing away from the wind produces less lift due to greater sweep. These lift changes produce a rolling moment that tends to right the aircraft. The magnitude of the lateral stability contribution depends upon the amount of sweepback and the lift coefficient of the wing with the stability contribution increasing with increasing amounts of sweepback and lift coefficients.

As might have been anticipated, there are also disadvantages associated with sweepback that are important both from the standpoint of aircraft design and flight operations. When sweepback is combined with taper (which it almost always is on modern aircraft), there is an extremely powerful tendency for the wingtip to stall first. A serious reduction in lateral control effectiveness, little stall warning, and a forward shift of the center of pressure (tending to exacerbate the stall) make this pattern of stall very undesirable. Taper produces higher local lift coefficients toward the tip, and one of the effects of sweepback is very similar. All outboard wing sections on a swept-back or tapered wing are affected by upwash from the preceding inboard sections. Additionally, a tendency to develop strong spanwise flow of the boundary layer toward the tip when the wing is at high-lift coefficients exists. This spanwise flow produces a relatively low energy boundary layer near the tip, which is easily separated.

The combined effect of taper and sweepback is illustrated by the flow patterns shown in Fig. 2-11. Design requirements for high-speed performance might dictate high sweepback while structural efficiency and weight considerations might demand a highly tapered wing planform. When these factors are controlling, the wing

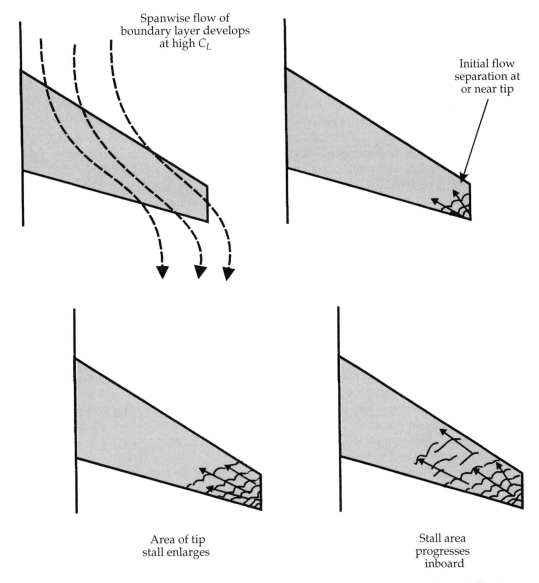

Fig. 2-11. The typical stall pattern for a swept wing begins at the tip and progresses inboard. The spanwise flow of the boundary layer at high lift coefficients contributes to this characteristic.

might require extensive aerodynamic tailoring to provide suitable stall characteristics and the sought-after reduction of drag in cruise conditions. Tip washout, flow fences, variation of camber throughout the span, and leading-edge devices are all methods employed to modify or limit the stall pattern induced by the tapered swept-back design.

The use of sweepback will reduce the lift curve and the subsonic maximum lift coefficient. This is a subsonic-only effect since we have already seen that sweepback will improve transonic maneuvering capability. The reduction in low-speed maximum lift coefficient has important implications. Stall speeds will increase unacceptably if wing loading is not reduced, and maneuverability will be decreased. Decreasing the wing loading, however, will increase the wing surface area and might negate some of the benefits of sweepback in the transonic flight regime. Because increased speed and transonic performance seem to be of prime importance, certain increase in stalling speeds with the consequently higher takeoff and landing speeds will be tolerated. While reduction in the lift-curve slope might provide a smoother ride, the consequent reduction in sensitivity to changes in angle of attack has certain undesirable drawbacks for low-speed flight. The reduced wing-lift curve results in increased angles of attack required for maximum lift, complicates landing gear design, and inhibits cockpit visibility. This effect can be readily observed during the final landing approach of a swept-wing jet. The higher the angle of sweepback, the higher the deck angle tends to be on landing. Leading- and trailing-edge devices can only partially overcome this effect. A lower lift-curve slope will also degrade the contribution to stability of a given tail-surface area.

Because the lift curve of a swept wing does not have a sharp break at the stall point, many pilots become somewhat complacent when transitioning into swept-wing aircraft. While it is true that the swept-wing aircraft does not show a clean stall break, it is also true that sweeping the wings decreases the aspect ratio, and this produces a large increase in induced-drag coefficient. This can be particularly critical during takeoffs and landings.

Use of sweepback will reduce the effectiveness of trailing-edge control surfaces and high-lift devices. A single slotted flap applied over the inboard 60 percent span to both a tapered and a 35° swept-back wing will have widely different effectiveness. The flap on the straight wing will increase the lift coefficient by approximately 50 percent, while the same flap applied to the swept wing will only produce a 20 percent increase. This can be partially overcome by unsweeping the flap hinge line and/or applying leading-edge high-lift devices such as slot or slats or employing some method of boundary layer control.

As previously noted, sweepback improves the lateral stability of the aircraft by providing stable rolling moments with sideslip. Again, the contribution to stability will be dependent upon the amount of sweepback and the wing's lift coefficient. Larger sweepback and high lift coefficients produce larger contributions to lateral stability. An excess of stability, however, inhibits controllability. An excess of lateral stability can aggravate any Dutch roll characteristics the aircraft might have and produce marginal control during crosswind takeoff and landing.[10]

Finally, the structural complexity and aeroelastic problems created by sweepback are of great importance. As shown in Fig. 2-12, a swept wing has a longer structural span for its aerodynamic span than a comparable straight wing of the same area, aspect ratio, and taper. This increases the wing structural weight because of the requirement for materials that can withstand greater bending and have greater shear strength. Also, large twisting loads are created at the wing root and the carry-through area.

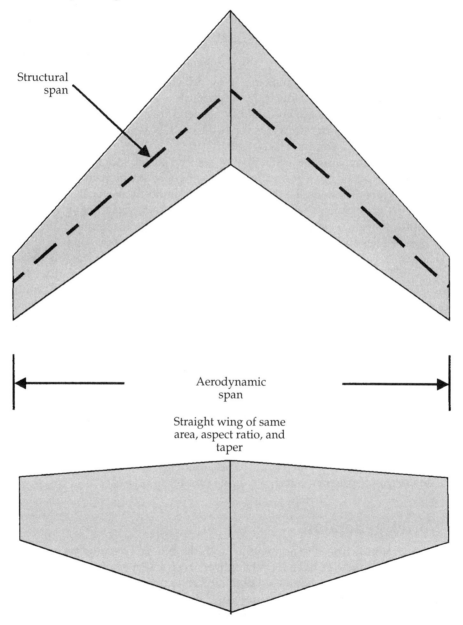

Fig. 2-12. A swept wing and a straight wing with the same aspect ratio, area, and taper will have different structural spans and different structural requirements.

Airfoil sections

Airfoils used for high subsonic flight must be somewhat different from those on slower aircraft. It should be obvious that the airfoil selected should have high critical Mach numbers since this number defines the lower limit for shock wave formation and subsequent force divergence. A very thin airfoil would be ideal for transonic flight since the air would be speeded up less when flowing over the top surface allowing higher airspeeds to be reached before the flow reaches sonic speed. Problems associated with thin wings, however, include higher takeoff and landing speeds due to the lower C_{LMAX} attained by these airfoils.

The thinness of the wing makes it very difficult to attach the leading-edge high-lift devices necessary to lower these speeds. It is also far more challenging to design the structural strength and rigidity needed into a thin wing. A final problem is the lack of room in thin wings for sufficient fuel tank volume to provide a reasonable range for a commercial aircraft.

The laminar flow airfoil was the first high-speed airfoil. Moving the maximum thickness of the airfoil back from the 25-percent chord position to 40–50 percent chord does reduce drag and raise the critical Mach number; however, the shock wave that develops on a laminar flow airfoil develops in the adverse pressure gradient area. Airflow that is slowing up in this region has a tendency to separate more easily. Many business jets and airliners today still use lower thickness laminar-flow airfoils.

The supercritical airfoil shown in Fig. 2-13 was designed by the scientists at NASA, chiefly Dr. Richard Whitcomb, with the aid of advanced computer techniques. This airfoil has a critical Mach number very close to Mach 1. The maximum thickness of this airfoil is very far back toward the trailing edge. Additionally, the upper surface of this airfoil has very little curvature. This prevents the rapid acceleration of air over the surface and eliminates the large pressure peak characteristic of low-speed airfoils. Instead, the pressure buildup over the upper surface of the supercritical airfoil is fairly flat but covers much of the surface.

At critical Mach number, the flow is only slightly supersonic, but widely distributed with the supersonic flow beginning near the leading edge. The flow is gradually decelerated near the trailing edge to subsonic speed. This gradual deceleration discourages the formation of a shock wave and results in low drag even at speeds approaching Mach 1. As can be seen from the illustration, this airfoil has sufficient thickness to provide for structural strength, the attachment of leading edge devices, and sufficient internal volume for fuel storage.

Vortex generators

Vortex generators are used to add energy to the boundary layer and delay airflow separation. These devices have found application on all types of aircraft and have become very popular on general aviation aircraft as a means of lowering stalling and minimum-control speeds and providing shorter takeoff and landing speeds and higher maximum allowable gross weights.

The same general principles apply to the regime of high-speed flight. The shock-wave induced separation occurs because the boundary layer does not have

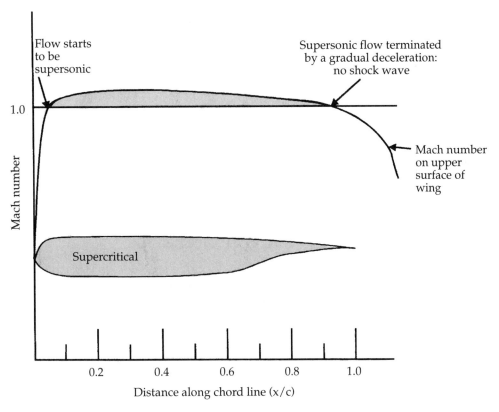

Fig. 2-13. The velocity of the airflow over a supercritical airfoil shows little variation with chord position and decelerates gradually toward the trailing edge. No shock wave is formed.

sufficient energy to overcome the adverse pressure gradient through the shock wave. The vortex generators[11] are simply small airfoil surfaces that protrude vertically into the airstream. A vortex is formed at the tip of each winglike protrusion that will act to mix the higher kinetic energy air from outside the boundary layer with the slower, lower-energy air of the boundary layer and delay separation. If a normal shock wave does form, the vortex generators are effective in breaking it up and thereby delaying separation. The benefits derived from vortex generators outweigh the additional drag incurred by their installation.

Vortex generators are normally installed in front of control surfaces such as ailerons, flaps, rudders, and elevators to prevent formation of shock waves in front of the hinge points of these control surfaces. If shock waves form in this position, the control surfaces can be rendered ineffective and aircraft control is seriously jeopardized.

All-movable control surfaces

Conventional control surfaces with fixed and movable portions such as the horizontal stabilizer-elevator combination suffer problems produced by shock wave formation in the transonic regime. A shock wave tends to form at the junction of these surfaces preventing the elevator from having any influence on the stabilizer ahead of it. This seriously compromises the pitch control of the aircraft. This could, as we have seen, be disastrous in a tuck-under or tuck-up situation.

This problem first came to light in the Korean-War era when fighters were first reaching or exceeding the speed of sound. Good old Yankee ingenuity won out and the all-movable tail surface was developed. This surface eliminates the stabilizer-elevator junction and the consequent shock wave by moving the whole stabilator to control pitch. (The stabilator is also seen on the Piper Cherokee series of single-engine aircraft, although it is highly unlikely that these aircraft suffer any transonic flight problems.) Because of this innovation in flight control, the American F-86 Sabres had an advantage over the faster and more powerful MiG-15s, which had conventional controls.

STABILITY AND CONTROL

Jet-aircraft stability ranges from the extremely docile Cessna Citation 500 series of aircraft that can be easily hand flown at flight-level altitudes through the Learjet series that have altitude and airspeed limitations with the autopilot inoperative to the latest and greatest of the United States military's machines, which are totally uncontrollable without all their automatic flight computers working.

Discussions in this section will be confined to civil aircraft and to some of the stability and control problems and solutions particular to high-altitude and/or high-speed flight or induced at lower speeds by the aircraft configurations adapted to high speeds and altitudes.

Dutch roll

Dutch roll is a problem at low speeds, but is a result of aircraft design features that allow higher M_{CRIT} speeds. Dutch roll is a coupled directional and lateral oscillation that occurs when the dihedral effect is very strong in comparison to directional stability. Recall that a swept wing with a high lift coefficient will have a very strong dihedral effect. This will increase stability, but will decrease controllability.

If an airplane is in a sideslip to the left, it will yaw to the right. The right wing will develop more lift, and the aircraft will roll to the left. If not controlled, the up-going right wing will cause a sideslip to the left and the whole process will repeat. This process can be provoked by turbulence, poor rudder control, or overcontrolling by a ham-handed pilot (We've talked about this person before, haven't we?) This can happen at any speed, but since the swept wing is less efficient at low speeds it usually occurs as a result of the pilot having a heavy hand or foot while maneuvering. Because this is an uncoordinated maneuver, it can be very upsetting to passengers and can be lethal in some aircraft (the early Model 23 Learjets in particular).

The cure for the Dutch-rolling tendency is an operable yaw damper. The yaw-computer accelerometers and gyros sense the yawing motion and reposition the rudder to stop the yaw. In some aircraft, this is coupled to aileron control also to correct for the bank left after the yaw is stopped. Dutch roll can be stopped manually by holding the rudder pedals steady in the neutral position and recovering from the resulting bank. These maneuvers should only be practiced in appropriate simulators. Some aircraft, such as some of the early Lears, must have yaw damp engaged from liftoff to touchdown.

Roll control

One of the adverse effects of the swept wing can be aileron ineffectiveness due to the spanwise flow induced by this planform. Additionally, any shock waves present will deteriorate aileron effectiveness. A solution to this problem has been the use of flight spoilers or spoilerons to aid in roll control. In some aircraft, the spoilerons are used only in low-speed mode or with the extension of flaps. In others, spoilers augment aileron action at all speeds. When used, the spoilerons move in a one-to-one relationship with the ailerons with the spoiler being extended on the wing with the up aileron and faired on the wing with the down aileron.

Inboard ailerons are used on some models of aircraft to overcome problems in aileron effectiveness caused by sweepback. In these aircraft, only the inboard set of ailerons is used in high-speed flight; both sets are used for low-speed operations.

SPEED CONTROL

Speed control addresses both ends of the speed envelope for jet aircraft. Precise speed control is important for the proper operation of any aircraft, but it becomes imperative in jets. Exceeding the speed limitations at either end of the spectrum in jets is likely to result in some very dicey moments for the crew and passengers and can result in structural failure at the extremes. Stick shakers, nudgers, pushers, pullers, Mach overspeed warnings, and angle-of- attack indicators are all installed on aircraft to prevent the pilot from exceeding limitations.

High-speed devices

Mach trim The Mach-trim system is an automatic trim system that enhances the longitudinal stability of the aircraft during accelerations/decelerations at high Mach numbers to compensate for Mach-tuck tendencies. The Mach-trim system normally uses the autopilot pitch-trim servos and receives input from an air-data sensor (in the pitot static system) to its computer. The Mach-trim computer will integrate the information from the air-data sensors and will automatically become active at a predetermined Mach number. (This will obviously vary with aircraft make and model.) The Mach-trim system will then function to automatically apply pitch trim in the proper direction (noseup trim for increasing Mach, nosedown trim for decreasing Mach) through the primary pitch-trim motor.

Mach overspeed warning Mach overspeed warning systems consist of an aural warning and a stick puller. Input is received from the pitot static system through the air-data computer. If M_{MO} is exceeded, an aural warning will sound (this is sometimes referred to as a chirper or a clacker, depending upon its sound) and the stick puller will be activated to rotate the nose up and slow the aircraft. Most systems use the autopilot pitch servos to accomplish this task.

Low-speed warning devices

Stick shakers, nudgers, and pushers These devices work with signals from the angle-of-attack system. The angle-of-attack system consists of a sensor that protrudes from the fuselage into and aligns itself with the average free airstream in the region of the leading edge of the wing. The sensor provides information to the angle of attack/stall computer that integrates this signal with flap position signals and rate information to detect approaching stall angle of attack. Aneroid sensors bias the system for the changes in stalling angle of attack normally occurring at high altitudes. The system uses stick-shaker motors located on each control column to alert the pilot that a stall condition is approaching.

Should the angle of attack continue to increase beyond shaker activation, a nudger (on some aircraft) using the pitch servos will apply a gentle, pulsating forward "nudge" to the control column. If neither of these warnings have convinced the pilot to stop the increase in the angle of attack, the stick pusher will take over and using the pitch servos will apply an elevator down motion causing an abrupt forward movement of the control column.

These stall warning and avoidance devices are necessitated by the less than desirable stall characteristics exhibited by swept-wing aircraft, and especially those with T-tails. As has been pointed out, utilizing a swept-back wing has the advantage of raising the critical Mach number. The disadvantages, however, include control problems near the stalling regime. Because the tip stalls first on a swept wing, the effect is to move the center of lift forward and cause a noseup pitching moment, further exacerbating the stall. The final result is a reverse of the normal positive longitudinal stability. At some critical angle of attack, the nose begins to pitch up itself, and elevator authority might not be sufficient to prevent this occurrence.

EMPENNAGE

T-tails are efficient in high-speed flight. They are above the area of disturbed air from the wing roots and fuselage. The T-tail can consequently be smaller in area and operate at smaller angles of attack and create far less drag. This high tail significantly reduces drag, but it can also cause significant problems near the stall regime.

Conventional tails operating near the stall regime are increasingly immersed in disturbed air from the wing. This reduces the horizontal surface's effectiveness and induces a nosedown pitching moment—or at least limits the tail's pitchup authority. The T-tail, however, remains out of this disturbed flow until very high angles of attack and therefore retains its full pitchup authority, which in some cases makes it possible to pull the nose up far beyond the normal stalling point. After the nose comes up, a critically high angle of attack develops and the tail finally dips into the low-energy wake of disturbed air from the fuselage and wing roots. Just when elevator effectiveness is needed to counter the extreme noseup pitch, it is no longer available.

The stabilator—that all-movable control surface that is so effective in preventing shock wave formation—now becomes a liability. The small elevator on such installations is more of a trim tab designed to effectuate rather small, short-term angle-of-attack adjustments. Large and sustained adjustments are reserved for the stabilator trim. The elevator just might not be large enough to make the needed pitch changes.

Aft mounted engines will contribute negatively to this whole scenario. While they have many positive advantages including lower V_{MC}, aerodynamically clean wings, lowering of cabin noise, and reduction of danger of foreign-object damage, these advantages do not come without some trade-off. Aft-mounted engines obviously shift the center of gravity aft. In order to keep the center of lift aft of the center of gravity, the wings must also be moved aft. This leaves a long, overhanging fuselage in front of the wing. The inertial forces associated with the large mass of the fuselage tend to generate an overswing beyond the desired pitch adjustment that can induce or aggravate the pitchup that can lead to a deep stall. Additionally, far past the stalling point, when the wings have shed most of their lift, that large, forward fuselage—which is also a nice fat airfoil—will continue to generate lift with a noseup component, exacerbating an already bad situation.

The aircraft might well settle into a tail-low descent into the real estate, or the fuselage might eventually reach its stalling angle of attack and fall through to a near-level attitude. The aircraft now stabilizes into a flight path normally reserved for falling bricks and is in a deep stall. Because none of the flight controls were designed to work at a 90-degree angle of attack, control is impossible. This should shed some light on why aircraft manufacturers provide so many stall warning/stall avoidance devices on this class aircraft.

If all this sounds grim, it is; however, it is also reserved for the inattentive or ignorant pilot who is not aware of or knowingly exceeds aircraft limitations and warnings. The overconfident, "I can fly anything through anything without no fancy training" attitude has always been diametrically opposed to long life as a pilot. The end just comes sooner for these folks in jets.

ANGLE-OF-ATTACK SYSTEMS

The angle-of-attack system found on most jets is one of the most valuable and underutilized systems on the aircraft. Very few pilots, other than Naval aviators, seem to use it to its fullest potential. Aircraft performance is based upon angle of attack. An airfoil will always stall at the same angle of attack, even though that might come at highly varying airspeeds. This subject will be covered more fully in chapter 8, but suffice it to say that observing changes in indicated angle of attack can give the pilot advance warning of many potential problems.

Endnotes

1. The sound barrier was broken by Gen. Yeager on October 14, 1947, over Muroc (later Edwards) Air Force Base in California's Mojave Desert. He achieved a speed of 697 mph (550 knots) in the Bell X-1 research aircraft named *Glamorous Glennis*, after his wife.

2. Surely you have heard the question "If a tree falls in the woods and there is no one around to hear it, does it make a sound?" I leave that to you to answer for yourself.

3. It should be noted that shock waves can and do form on surfaces other than wings, but we are most interested in the effect on wings.

4. Heat has been called the graveyard of energy. Any time there is heat dissipation, there is energy loss.

5. There have been accidents in certain early models of the Learjets where the accident investigators theorized that the crash had been the ultimate effect of a jet upset caused by flying faster than M_{CRIT}. One thing that caused this determination was the tripped condition of the Mach/overspeed warning and/or stick-puller circuit breakers.

6. This might seem to negate my previous statements that jets were not difficult to fly. For these compressibility effects to become apparent, the aircraft limitations must be exceeded. Modern aircraft have systems and procedures to ensure that this does not take place. This book was not written for those who choose to deliberately exceed published limitations.

7. True, I don't know why a pilot would be hand-flying in cruise at this altitude. It is at the very least a fatiguing exercise. Many aircraft limitations make use of the autopilot mandatory above specified altitudes and/or Mach numbers, but it is possible to be ham-handed with an autopilot, also.

8. The *Airline Transport Pilot and Type Rating Practical Test Standards* require in "Area of Operations IV—Inflight Maneuvers, Task E: Recover from Specific Flight Characteristics" that the applicant "1. Exhibits adequate knowledge of recovery from specific flight characteristics (as appropriate to the airplane); and 2. Uses proper technique to enter into, operate within, and recover from specific flight characteristics." Even though first officers are not usually typed in the aircraft, they will usually receive the same training as required for a type rating. Use of sophisticated flight simulators has made training and testing in this area practical and safe.

9. This is not an insubstantial consideration in passenger aircraft. Perhaps you have noticed as I have that it seems that you could be totally disoriented enroute and be totally lacking in situational awareness, but if the ride is smooth and the landing a roll-on, the passengers think you are the world's greatest pilot.

10. This is one of the reasons you might see many jet airliners crabbed in a crosswind until the very last moment when they are "kicked" straight. Landing techniques that work in slow, straightwing aircraft aren't always appropriate in faster, swept-wing equipment.

11. Some models of the Learjet use boundary layer energizers—BLEs—instead of vortex generators to accomplish the same purpose. The BLEs are incorporated into the "soft-flight" wing modification and are claimed to be more efficient than vortex generators. Although the Learjet is authorized for flight with up to six vortex generators missing (no more than three on either wing), if any BLEs are missing, a reduction in M_{MO} from M 0.83 to M 0.78 must be observed.

3

High-altitude physiology

I AM ALWAYS AMAZED WHEN I ASK A CLASS OF EXPERIENCED JET PILOTS undergoing recurrent training how many have been through the FAA's physiological training program. Usually less than half will have taken advantage of this valuable opportunity. We spend many hours in our carefully conditioned cockpits, only inches from an environment that is incompatible with life. We sit blissfully unaware of the perils that exist just the other side of the plexiglass. We sit in shirt-sleeves, comfortably oblivious to the frigid and rare atmosphere through which our fragile shell moves.

How many consider the temperature outside or the time available to react should the carefully maintained environment fail? Jet aircraft are designed to operate efficiently at high altitudes, but the human body is not. In all of flying, it is important to remember that humans are land animals, evolved to exist comfortably close to sea level at a maximum speed (and that for only very short sprints) of little more than 15 mph. Anything else is a foreign and potentially lethal environment. Risk exists whenever a human operates above the altitude of acclimatization, which is the altitude at which she or he resides. No matter how you perceive your performance (and despite all the bravado and tough war stories), the body will still respond to the atmosphere in which it is operating and be affected by gas concentrations and ambient pressures.

This chapter will explain the atmosphere and the various laws that govern the behavior of gases. We will explore the composition of the atmosphere and the effects upon health and performance, the effects of temperature and pressure, and the various physiological zones of the atmosphere. An understanding of the makeup of the atmosphere from a physiological point of view will help the pilot recognize and understand the potential dangers and subtle changes induced by atmospheric variables. Only with an awareness of the atmosphere's impact will a pilot have a true respect for human vulnerability.

COMPOSITION OF THE ATMOSPHERE

The atmosphere is an envelope of air that surrounds and rotates with the Earth to an altitude of about 25,000 miles. This envelope is constantly changing, as every weather briefing will reaffirm, and temperatures and pressures fluctuate from day to day and from region to region. From a biological point of view, however, the atmosphere can be considered a constant in relation to its effect on the human physiology.

Although it is often said that the air is "thin" at altitude and lacks the oxygen necessary to sustain life, the actual composition of the atmosphere remains constant throughout the altitude range. The percentage of oxygen in the air is constant at 21 percent, but the actual number of oxygen molecules per unit volume of air decreases with pressure and consequently with altitude. There will always be a constant 21 percent oxygen available, but it is analogous to the difference between 21 percent of one dollar and 21 percent of one cent. The percentage is the same, but the value is vastly different. The remaining 79 percent of the composition of the atmosphere is principally nitrogen (78 percent) with carbon dioxide (0.3 percent), inert gases (1 percent), and water vapor.

Physical characteristics of the atmosphere
Pressure

Atmospheric pressure is really just the weight of all the air molecules in the atmosphere that are located above the point at which the measurement is made. Because there are fewer molecules in the column above the measurement point at higher altitudes, it can be readily seen that atmospheric pressure decreases with increasing altitude. Because air is compressible, the atmosphere will be denser near the surface (the bottom of the column) and increases of pressure are greater nearer the surface. Conversely, the pressure decreases more rapidly the higher in the atmospheric column that one rises. The greatest density change occurs between sea level and 5000 feet; therefore, the problems associated with pressure and density change must be considered even in pressurized aircraft.[1]

A variety of atmospheric properties will change this pressure besides its weight. Seasonal temperature changes, weather systems, latitude and longitude, and time of day all affect atmospheric pressure. The International Standard Atmosphere (ISA) is a baseline from which to work. This standard establishes a mean atmospheric pressure of 29.92 inches (760 mm) of mercury at a temperature of 15°C (59°F) in dry air at sea level. This is also expressed as 14.7 pounds per square inch (PSI) or 1013.2 millibars (MB) at the same temperature.

Temperature

The surface of the Earth is warmed by solar radiation that is then reflected back into the atmosphere. This reflected radiation does little to heat the atmosphere and consequently the temperature of the atmosphere decreases with increasing altitude until the level of the tropopause is reached (about 35,000 feet) at which the temperature remains relatively constant. The decrease in temperature, or *lapse rate*, is for dry air and is 3.56°F (1.98°C) per thousand feet of altitude.

Table 3-1 shows the standard decrease in pressure with increasing altitude and the standard temperatures associated with those altitudes.

Table 3-1. International standard atmosphere

Altitude feet	Pressure Inches Hg	mm/Hg	PSI	Temperature °C	°F
Sea level	29.92	760.0	14.7	15.0	59.0
10,000	20.58	522.6	10.1	−4.8	23.3
18,000[1]	14.94	379.4	7.4	−20.7	−5.3
20,000	13.75	49.1	6.8	−24.6	−12.3
25,000	11.10	281.8	5.5	−34.5	−30.2
30,000	8.90	225.6	4.4	−44.4	−48.0
34,000[2]	7.40	187.4	3.6	−52.4	−62.3
36,089[3]	6.68	169.8	3.3	−56.6	−69.7
40,000	5.54	140.7	2.7	−56.6	−69.7
45,000	4.34	110.6	2.1	−56.6	−69.7
50,000	3.43	87.0	1.7	−56.6	−69.7
60,000	2.12	53.8	1.0	−56.6	−69.7

1. ½ atmosphere
2. ¼ atmosphere
3. Tropopause

Gas laws

The mixture of gases in the atmosphere is subject to several laws of physics governing the behavior of gases. An understanding of these laws will help in the understanding of the effects of altitude and these gases in the body.

Dalton's law Dalton's law states that the total pressure of any mixture of gases (with the temperature and volume held constant) is the sum of the individual pressures (also called partial pressure) each gas in the mixture would exert if each gas alone occupied the total volume. Furthermore, the partial pressure of each gas is proportional to that gas's percentage of the total mixture. Because the percentage of oxygen in the atmosphere remains constant at 21 percent, Dalton's law will allow us to calculate the partial pressure of the oxygen in the atmosphere at any given altitude. Table 3-2 shows the partial pressure of oxygen in millimeters of mercury (mm/Hg) at various altitudes.

The human body is affected by the pressure of the gases in the atmosphere. In particular, the partial pressure of oxygen (and to a lesser extent other gases) available in the surrounding air is of vital importance in determining the onset and severity of hypoxia.

Henry's law Henry's law states that the amount of gas dissolved in a solution is directly proportional to the partial pressure of that gas over the solution. This law can be most clearly illustrated with a bottle of carbonated liquid. When the bottle is uncapped, the carbon dioxide (CO_2) in solution in the mixture will slowly diffuse to the atmosphere until the pressure of CO_2 in the liquid equals the

Table 3-2. Partial pressure of oxygen varying by altitude

Altitude feet	Total pressure (mm/Hg)	Oxygen pressure (mm/Hg)
Sea level	760.0	159.6
5,000	632.3	132.8
8,000	564.5	118.6
10,000	522.6	109.8
15,000	428.9	90.1
20,000	349.1	73.3
25,000	281.8	55.2
30,000	225.6	47.4
35,000	178.8	37.6
40,000	140.7	29.5
45,000	110.6	23.2
50,000	87.0	18.3
60,000	53.8	11.3

pressure of CO_2 in the surrounding air. The beverage will then be "flat." A bottle of soda opened in an unpressurized aircraft at 10,000 feet will foam and overflow the container. The opposite will happen with soda opened at pressures greater than one atmosphere. A champagne cork would not pop in a diving bathysphere pressurized for deep ocean exploration (Fig. 3-1).

Boyle's law Boyle states that the volume of a gas is inversely proportional to the pressure on the gas, providing the temperature remains constant according to the formula $V1/V2 = P2/P1$ where $V1$ and $P1$ are the initial volume and pressure and $V2$ and P are the final volume and pressure. Stated more simply, a gas will expand when the pressure on it is decreased. This law holds true for all gases, and in particular for those trapped within body cavities. (Although, within the body, the value of 47 mm Hg must be subtracted from each pressure to account for the presence of water vapor in the body.) A volume of gas at sea-level pressure will expand to approximately twice its original volume at 18,000 feet, nearly nine times its original volume at 50,000 feet. (See Fig. 3-2.)

Graham's law Graham's law states that a gas of higher pressure exerts a force toward a region of lower pressure. In the presence of a permeable or semipermeable membrane separating these regions of gases of unequal pressures, the gas will diffuse from the region of lower pressure to the region of higher pressure. This process will continue until the pressure of the gas is equal, or nearly equal, on both sides of the membrane. Graham's law holds true for all gases, and each gas behaves independently according to the pressure differential across the membrane for that gas. It is possible to have two or more gases in a solution diffusing in opposite directions across the same membrane and, in fact, this is just what happens to make oxygen transfer possible in the cells and tissues of the human body.

Fig. 3-1. The amount of gas dissolved in a solution is directly proportional to the partial pressure of that gas over the solution.

PHYSIOLOGICAL DIVISIONS
OF THE ATMOSPHERE
The fairly safe zone

The fairly safe zone of the atmosphere extends from sea level upwards to approximately 10,000 to 12,000 feet. It is in this zone that the human body can adapt to work and live. The key word here is adapted. A human adapted to life at sea level where the atmospheric pressure is 760 mm Hg is not going to be immediately efficient or even comfortable at 10,000 feet where the barometric pressure has dropped to 523 mm Hg. After a period of acclimatization, however, normal life and work activities are possible. Decompression sickness and altitude sickness are rare at these altitudes, although they do occur.

In this lower region of the atmosphere, there is sufficient partial pressure of oxygen to sustain an otherwise healthy pilot at a reasonable level of consciousness and efficiency. Peak efficiency will not be maintained, however, even in the healthiest and most robust person who has not spent a period of acclimatization. Hypoxia and decreased efficiency and awareness are possible, even probable, any time a human is exposed to altitudes above that to which he or she has adapted.

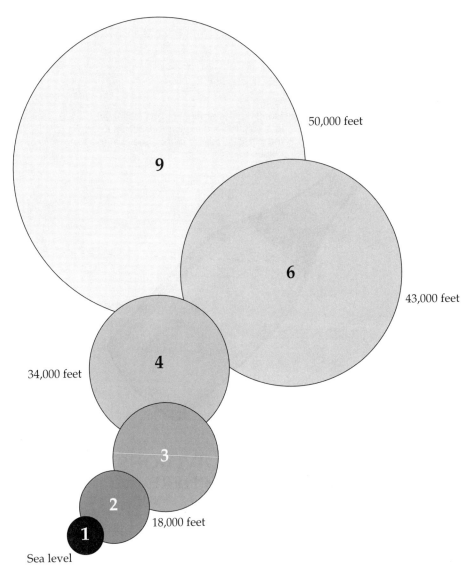

50,000 feet

9

6

43,000 feet

34,000 feet

4

3

2

18,000 feet

1

Sea level

Fig. 3-2. Expansion of gases with increasing altitude—and consequent decreasing ambient pressure—results in a volume of gas at 50,000 that is nine times that of the same gas at sea level.

The physiologically deficient zone

The physiologically deficient zone of the atmosphere extends from the top of the fairly safe physiological zone to about 50,000 feet. Above 50,000 feet, life is not possible without pressure suits or pressure vessels. This area above the physiologically deficient zone is known as the impossible zone or the space equivalent division. The body responds to ambient conditions here as though it were in deep space. At about 63,000 feet, also called Armstrong's line, body fluids such as blood and cerebral spinal fluid boil causing air embolisms in the brain and other body organs and instant death. The boiling point of water at the low ambient pressure at 63,000 feet (about 46.5 mm Hg or 0.9005 PSI) is 98.6°F, or that considered normal human body temperature. Remember that the boiling point of any liquid is that point at which it changes from a liquid to a gas.

Operation in the physiologically deficient zone of the atmosphere poses an increased risk of hypoxia, evolved gas disorders, and trapped gas disorders. Although all turbine aircraft and pressurized piston aircraft routinely operate in this zone, we humans are out of our normal element here, dependent upon the integrity of our fragile cocoon, and at risk. In the next section, we will explore the effects of reduced atmospheric pressure on the human body.

EFFECTS OF ALTITUDE ON THE HUMAN BODY

When we talk about the effects of altitude upon the human body and altitude sicknesses, we tend to think in terms of "high altitude" and classify that as somewhere in the flight levels. Nothing could be further from the truth. As stated often before, the human body will be adversely affected by prolonged exposure to any altitude above that at which it has adapted to live.

A resident of coastal California will not perform nearly as efficiently at any task in Denver, where the ground is a mile high, as will a native Denverite, but most would not consider being on the ground at Denver as being at "high altitude." Your body and brain, however, will have a different perspective on the matter. The discussion that follows applies to all pilots: jet jocks, helicopter pilots, recreational pilots, and hang glider aficionados.

OXYGEN USE IN THE HUMAN BODY

It seems logical to start our discussion of the effects of altitude on the human body by first discussing how the body normally acquires, transports, and uses oxygen. The importance and application of the gas laws discussed above will now become apparent. We are all aware that oxygen is necessary to sustain combustion or oxidation. It is necessary in the human body for the same reasons—to support the oxidation of fuels needed to provide energy for life.

Transport of oxygen

Very little of the oxygen carried by the blood is carried in dissolved form in the plasma. Most of the oxygen, almost 98 percent of that transported, is transported by the hemoglobin molecules in the red blood cells. The ability of hemoglobin to combine with and transport oxygen is dependent upon the pressure of oxygen in the surrounding environment. Higher pressures of oxygen enable the hemoglobin to take up larger quantities of oxygen. Lower oxygen pressures will result in an increasing tendency by the hemoglobin to give up oxygen.

This variable combining characteristic is what allows the blood to acquire oxygen in the lungs and transport it to the tissues where it is given up. This characteristic of the hemoglobin also results in what is known as the *oxygen dissociation curve*. While oxygen pressure decreases in a straight line with altitude, the ability of the hemoglobin to hold oxygen follows a much different curve. (See Fig. 3-3.) Notice the dramatic change in the hemoglobin's ability to combine with oxygen that occurs in the low 20s.

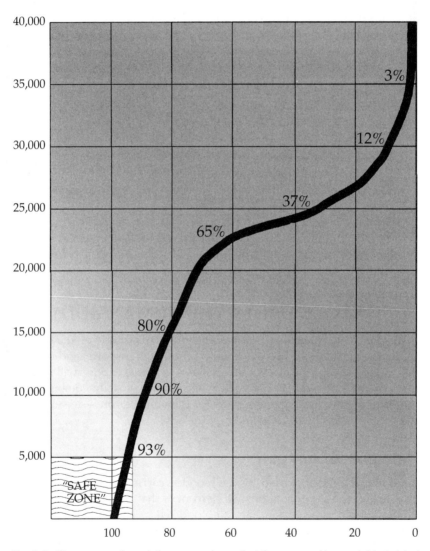

Fig. 3-3. The oxygen dissociation curve shows that the power of hemoglobin to bind oxygen is directly related to partial pressure of the oxygen and declines rapidly between 20,000 and 25,000 feet.

Air entering the lungs at sea level enters at a pressure of 760 mm Hg. As discussed before, this results in a partial pressure of oxygen in sea-level air of about 160 mm Hg. The blood flowing through the lungs is not, however, exposed to atmospheric air but rather to *alveolar air*, the air mixture contained in the tiny air sacks of the lungs, which is only 14 percent oxygen. (This is due to the addition of water vapor to the inspired air plus the carbon dioxide that has diffused from the blood returning from the tissues.) The partial pressure of oxygen in alveolar air is then 14 percent of 760 mm Hg or 106.4 mm Hg. Carbon dioxide, which is 5.5 percent of alveolar air (as contrasted to less than 1 percent in the atmosphere), exerts a pressure of 41.8 mm Hg.

The hemoglobin in the blood returning from the tissues carries oxygen, which exerts a pressure of 40 mm Hg. Graham's law governs the diffusion of oxygen from the higher pressure of the alveolar air to the blood and the diffusion of carbon dioxide from the blood to the alveolar sacks. The opposite transfer takes place when the oxygen-rich blood reaches the tissues that carry oxygen at an average pressure of 20 mm Hg. This lower pressure will allow the hemoglobin to release oxygen that will then diffuse into the tissues. (See Fig. 3-4.) Concurrently, carbon dioxide is diffusing from the tissues into the blood. (An average pressure for CO in the tissues is 50 mm Hg.; however, this is dependent upon the activity level of the tissue.)

In a normal, healthy individual, sea-level pressure is sufficient to cause the arterial blood leaving the lungs to be almost totally (97 percent) saturated with oxygen. At 10,000 feet, this saturation has dropped to almost 90 percent, which is still sufficient for nearly all usual life functions. An oxygen saturation of 93 percent is considered by medical personnel to be the low limit of normal functioning.

On top of Pike's Peak (about 14,500 feet and 438 mm Hg atmospheric pressure) the oxygen saturation has dropped to about 80 percent. Many people, if left in this rarefied air for some period, will develop mountain or altitude sickness: vertigo, nausea, weakness, hyperpnea (increased breathing), incoordination, slowed mental processes, dimmed vision, and increased heart rate. At 25,000 feet, the oxygen saturation is only 55 percent and consciousness is lost. (Note that the partial pressure of oxygen in alveolar air at 25,000 feet is 14 percent of 281.8 mm Hg or 39.5 mm Hg—slightly less than that normally found in venous blood returning from the tissues. Which way do you think the oxygen will diffuse at altitudes above 25,000 feet?)

TYPES OF HYPOXIA AND ITS EFFECTS

The effects of hypoxia upon flying skills and the symptoms of its onset are basically the same no matter what the cause of the hypoxia. It is useful, however, to look at some varying causes of this condition in order to be more alert to its potential onset in the presence of one or more of these factors.

Hypoxic hypoxia

Hypoxic hypoxia is also referred to by aviators as altitude hypoxia. This is the hypoxia that results when there is a lack of available oxygen or partial pressure of

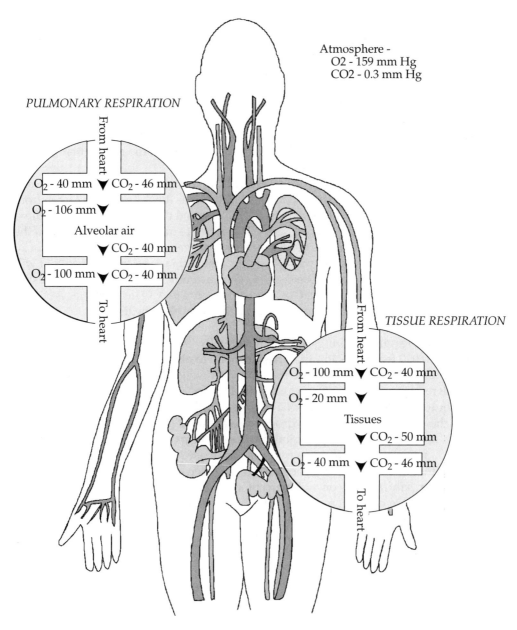

Atmosphere -
O2 - 159 mm Hg
CO2 - 0.3 mm Hg

PULMONARY RESPIRATION

From heart

O_2 - 40 mm ▼ CO_2 - 46 mm

O_2 - 106 mm ▼

Alveolar air

▼ CO_2 - 40 mm

O_2 - 100 mm ▼ CO_2 - 40 mm

To heart

TISSUE RESPIRATION

From heart

O_2 - 100 mm ▼ CO_2 - 40 mm

O_2 - 20 mm ▼

Tissues

▼ CO_2 - 50 mm

O_2 - 40 mm ▼ CO_2 - 46 mm

To heart

Fig. 3-4. Blood enters the pulmonary system depleted of oxygen and returns to the heart oxygen-rich. The opposite is true of blood going to the tissues; it leaves the heart rich in oxygen and returns depleted.

oxygen in the breathing air. This is the type hypoxia experienced when flying in an unpressurized cabin or when flying at altitude in a jet with a cabin pressurized to a cabin altitude above 5000 feet. Although strictly speaking we are somewhat hypoxic when operating even a few hundred feet above the altitude of acclimatization, this becomes most evident when flying unpressurized in the physiologically deficient zone. In reality, the symptoms of hypoxic hypoxia do not, in the absence of other contributing factors, become significant until about 5000 feet.

Hypoxic hypoxia occurs because there is a smaller and smaller pressure differential between the pressure of oxygen in the inspired air in the lungs and the pressure of the oxygen in the blood and tissues. Remember that the combining power of hemoglobin and oxygen is influenced by this pressure differential. The greater the differential, the more efficient the hemoglobin becomes. As this pressure differential lessens, it becomes harder and harder for the hemoglobin to pick up and transport the oxygen.

Hypemic hypoxia (anemic)

Hypemic hypoxia occurs whenever the blood's ability to carry oxygen is reduced, although there is sufficient oxygen at a sufficient pressure in the inspired air. A variety of conditions can cause this to happen.

Any condition that would cause a reduction in the number of healthy, functioning red blood cells (anemia or reduced production of red blood cells, blood loss, deformed blood cells, disease, etc.) will impair the blood's ability to supply the tissues with oxygen. Remember the old advertisements warning about "iron-poor blood"? Iron is the functional part of the hemoglobin molecule, and it is the iron that renders the hemoglobin absolutely indispensable for life. In addition to a reduction in the number of red blood cells available, anything that would interfere with the ability of hemoglobin to transport oxygen or anything that would displace the oxygen that is bound to the hemoglobin will affect the oxygen available to the cells.

The most common impairment to oxygen transport by the hemoglobin is carbon monoxide. Carbon monoxide combines with hemoglobin 200–300 times more readily than does oxygen and once bound is extremely hard to eliminate. Smokers will find that the carbon monoxide bound to their hemoglobin will lower their altitude for onset of hypoxic symptoms by 2000–3000 feet. This effect is not limited to smokers, however. Anyone exposed to a smoky atmosphere will suffer somewhat. (Remember this next time you volunteer to go along as a designated driver for a group of drinkers. Just sitting in that smoky bar for several hours is going to affect your performance the next day, even without alcohol and fatigue!) Other chemicals, among them sulfa drugs and nitrites (found in food preservatives), can have an adverse effect on the ability of hemoglobin to combine with and transport oxygen.

Histotoxic hypoxia

Histotoxic hypoxia is a disruption of cellular respiration. There might well be sufficient oxygen of sufficient pressure in the inspired air to fully saturate the blood and hemoglobin, but the cells expecting and needing the oxygen are unable to use it due to the presence or absorption of cell toxins. The most common toxin found at the cellular level that can cause this effect is alcohol. Although other toxins, notably cyanide and some narcotics, also can cause this disruption of cellular respiration, alcohol is by far the most common culprit.

Now, we are all aware of the hazards associated with alcohol and flying, and I'm not suggesting that any true professional would knowingly violate these rules and guidelines. Many pilots, however, might be impaired by alcohol at the cellular level and not be aware of the problem or its cause. Remember the iron-poor blood mentioned earlier? Be cautious of the "tonics" or "elixirs" offered as remedies. Carefully read the labels on any over-the-counter medications or nutritional supplements you propose to ingest. Although many more manufacturers are eliminating or reducing the alcohol content of the liquid medications, you might be surprised at the percentage of alcohol some still contain. One popular vitamin supplement for "iron-poor blood" contains 12 percent alcohol!

Stagnant hypoxia

Stagnant hypoxia results when the blood flow is compromised for any reason. This flow reduction can be due to impaired pumping efficiency by the heart, arteriosclerotic disease, and venous pooling of blood such as in varicose veins of the legs. Stagnant hypoxia can also occur at extremely low temperatures where blood supply to the extremities is decreased due to capillary constriction and rerouting. Stagnant refers to a diminishing of the blood flow to an area of the body, not necessarily complete stoppage, although this can certainly occur in extreme cases. Causes of this type of hypoxia peculiar to flying are the subjection to extreme G forces in a head-to-toe direction and prolonged periods of pressure breathing at altitude. Table 3-3 summarizes the various types and causes of hypoxia.

Table 3-3. Types of hypoxia and common causes

Type of hypoxia	Phase of respiration	Condition	Cause
Hypoxic Hypoxia	Ventilation	Reduction of PO_2 in inspired air	Altitude
			Asthma
			Breatholding
			Hypoventilation
			Insufficient PO_2 in mixture of breathing gases
			Malfunctioning oxygen delivery equipment at altitude
			Respiratory arrest
			Strangulation
Hypoxic Hypoxia	Ventilation	Reduction in area available for gas exchange	Congenital heart defects
			Drowning
			Emphysema
			Pneumonia
			Pneumothorax (air in chest cavity restricting inflation of lungs)

Type of hypoxia	Phase of respiration	Condition	Cause
			Pulmonary embolism (blood clot in lung)
Hypoxic Hypoxia	Diffusion	Barriers to diffusion of gases	Drowning
			Hyaline membrane disease
			Pneumonia
Hypemic Hypoxia	Transportation	Reduction in oxygen carrying capacity of blood or of hemoglobin	Anemias
			Chemicals (cyanide, carbon monoxide)
			Drugs (sulfa drugs, nitrites)
			Hemoglobin abnormalities
			Hemorrhage
Histotoxic Hypoxia	Utilization	Cellular poisoning or dysfunction	Alcohol
			Carbon Monoxide
			Cyanide
Stagnant Hypoxia	Transportation	Reduction in total body blood flow	Acceleration (G Forces)
			Heart failure
			Prolonged pressure breathing
			Pulmonary embolism
			Shock
Stagnant Hypoxia	Utilization	Local reduction of blood flow	Cerebral vascular accidents (stroke)
			Embolism by clot or gas bubbles
			Extremes in ambient temperature
			Hyperventilation
			Postural changes (prolonged sitting)
			Restrictive clothing, seat belts, etc.

Hypoxic effects

Hypoxia is an insidious and progressive condition and is almost undetectable by the pilot. You should always be aware that without supplemental oxygen at sufficient pressure you will gradually and progressively lapse into incompetence while maintaining an absolutely euphoric faith in your own ability.

As blood saturation of oxygen drops, there is a steady disruption of life functions. From a blood saturation of 93 percent, considered the low limit of normal functioning and where visual problems begin to occur, there is a rapid deterioration into unconsciousness with decreasing saturation.

Keep in mind, as mentioned earlier, that although the partial pressure of oxygen in the atmosphere decreases in a straight line with increasing altitude, the hemoglobin's ability to combine with this oxygen follows a much different and more deadly curve. This property of hemoglobin bears heavily on oxygen requirements at altitude. Let's look at some common hypoxic effects that an average, healthy individual can expect with increasing altitudes.

5000 feet This is considered by most to be a "low" altitude. The retina of the eye is more demanding of oxygen than any other organ of the body, even the brain itself, which demands 30 percent of the supply. At this "low" altitude, this extension of the brain will begin to suffer degradations in function that will be most noticeable at this point in night vision.

Instruments and maps are more easily misread during night flight at this altitude, and ground features and lights are more easily misinterpreted. It is always an eye-opening shock to my students when I bring them in over the Mojave Desert after a long flight at a cabin altitude of 8000 feet. After having them note the discernible features on the dark and trackless desert floor, I have them breathe 100-percent oxygen for a few minutes. Without exception, all are shocked and amazed to note the features that "jump" out of the blackness after a little O_2. Most have heard of this little demonstration before it actually takes place, but none are totally prepared for its dramatic effect. This level of hypoxia is extremely insidious because most pilots feel they are functioning at peak efficiency at this point. Extra vigilance by every pilot is necessary to prevent missing critical fixes on charts or misreading instruments.

10,000 feet Night vision is now degraded by 15–25 percent. The blood saturation has dropped to 90 percent, and your brain is receiving the absolute minimum supply of oxygen. This is the absolute highest altitude at which you should have any trust at all in your own performance, although your judgment is already severely compromised. Euphoria will prevent true self-assessment of your abilities. Physical hypoxic symptoms such as tingling and headache might not become apparent for 4 hours or more at this altitude, although judgment has long gone by the wayside. Above 10,000 feet, blood oxygen saturation and performance degrade steeply.

14,000 feet Blood oxygen saturation is down to a dangerous 85 percent. You will be increasingly disabled at this altitude. Vision will dim. You will experience serious degradation of judgment, memory, and thought. The impairment of judgment will leave you feeling just fine and confident in your performance, however. If hypoxia is not recognized and corrected at this stage of impairment, it is unlikely that it will be recognized. You are in serious danger.

16,000 feet This is only 2000 feet higher than the last assessment, but you will behave as though you had ingested a full load of gin and tonic. Your blood oxygen saturation will have dropped to 79 percent, and you will be seriously disabled. You will be euphoric, belligerent, disoriented, or perhaps all three. You will be irrational, unreliable, and dangerous. If you are alone, your chances of survival are decreasing rapidly.

18,000 feet At this altitude, you are incapable of any useful function although you might still feel great! Blood saturation has fallen to 71 percent and your brain is suffering. You will pass out in about 30 minutes.

20,000 feet If you have not already collapsed, it will not be long now. Five to 15 minutes is about the time of useful consciousness at this altitude, and prolonged exposure can result in death. Blood saturation has dropped to 71 percent.

25,000 feet Forget it. Blood saturation has now dropped to lethal levels. Time of useful consciousness is three to six minutes with death following not long after that. Above this altitude, suffering a rapid decompression might also result in a condition divers know as the bends and various other pressure-related maladies we will discuss shortly. Remember, this is only HALF as high as some modern civilian aircraft are certified to fly. Table 3-4 lists the times of useful consciousness (TUC) following decompression at increasing altitudes. Figure 3-5 summarizes the various medically recognized stages of hypoxia with the associated risk altitudes and blood oxygen saturations.

Table 3-4. Time of useful consciousness

Altitude (feet)	Time of Useful Consciousness (TUC)
15,000	30 minutes or more
18,000	20–30 minutes
22,000	5–10 minutes
25,000	3–5 minutes
28,000	2.5–3 minutes
30,000	1–2 minutes
35,000	30–60 **seconds**
40,000	15–20 **seconds**
45,000	9–15 **seconds**
50,000	6–9 **seconds**

FACTORS AFFECTING RESPONSE TO HYPOXIA

It is impossible to tell exactly when and in what manner hypoxic reactions will become apparent in an individual. Individual reactions to hypoxia vary greatly not only among people, but in the same person on a day-to-day basis due to differences in body chemistry, general health, and diet. Some of the determining factors are somewhat under the pilot's control and some are dictated by the flying environment itself.

Absolute altitude

This seems rather obvious. The severity of hypoxia will depend directly upon the absolute altitude of the environment in which the pilot is functioning. This might

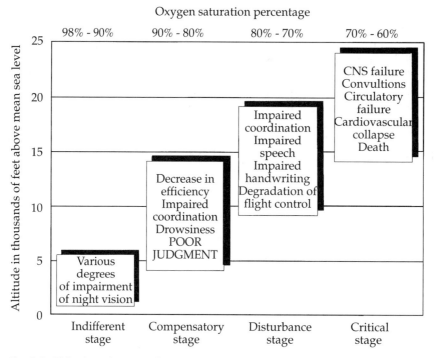

Fig. 3-5. This chart shows the four medically recognized stages of hypoxia along with their most common symptoms and altitudes of risk.

be the aircraft altitude in an unpressurized cabin or the cabin altitude in pressurized craft. As the altitude of the environment climbs and the partial pressure of the oxygen in that atmosphere drops, the risk of hypoxia rises. Acclimatization (to normal living altitudes) can help only to a limited extent. The Denver folks might have an advantage up to about 15,000 feet, but after that, everyone is more or less equally impaired. This factor is usually somewhat controlled by the pilot; however, mountains or weather can cause a climb to a previously unplanned altitude. This is of little consequence in a pressurized aircraft.

Acclimatization

One question I am often asked concerns the permanent residences in the Andes above 17,000 feet. How do these people remain conscious, much less do work? The answer is acclimatization. The Peruvians who live in these villages have extremely high red corpuscle counts and so have much more hemoglobin with which to transport oxygen. These residents also have much larger pulmonary ventilation volumes and increased cardiac output. Some acclimatization shown by this population is reversible; it will be lost with acclimatization to a lower altitude and consequently a higher atmospheric pressure. Other of their adaptations, however, seem to be permanent evolutionary responses to life at lowered atmospheric pressures and are not lost with residence at lower altitudes and are passed on genetically to subsequent generations.

The native Denverites we spoke of earlier will have more tolerance to altitude and become hypoxic later and at a higher altitude than the West Coast pilots because they are already partially acclimated to altitude by virtue of their residence in the Mile High City. When at 10,000 feet, the Denverites are only 5000 feet above their physiologically adapted altitude; the LA pilots would be 10,000 above theirs. (Then again, there are some who claim that those of us in LA are permanently in the ozone, but that is probably the topic of a different discussion.)

Rate of ascent

The quicker you (or your environment) climb, the more rapid the onset of hypoxic symptoms. The climbers who ascend Mt. Everest are well aware of this phenomenon. They spend several weeks in their climb, stopping at several intermediate altitudes to acclimate. An explosive decompression in an aircraft with the resultant rapid climb of the cabin altitude can reduce the time of useful consciousness to ⅓ to ½ of that listed in Table 3-4. A rapid ascent can cause the symptoms of hypoxia to quickly accumulate and incapacitate a pilot before awareness of the encroaching disability dawns on the dimming consciousness.

Time of exposure

Staying at 8000 feet for several hours (not uncommon in a jet at flight-level cruising altitudes) can cause the same symptoms and incapacitation as staying at a higher altitude for a shorter duration. The symptoms of hypoxia are cumulative and time related, but there is no reliable means to predict the exact relationship or effect. The only certainty is that the higher the altitude, the shorter the time of exposure before symptoms begin to occur. This, too, will vary on an individual basis.

Physical activity

Any physical activity will obviously cause the body to demand more oxygen for normal functioning. The muscles will rob the brain of the marginal amounts of oxygen available in the blood, and the time of onset of hypoxic symptoms will be shortened. Although not much physical activity is expended by pilots, the extra amount required to fly in turbulence or with a failed autopilot can dramatically reduce the already minimal oxygen supply to the brain and the retinas of the eyes. This factor is usually not under the pilot's control.

Temperature

The temperature in the cabin has a great effect on an individual's tolerance for and response to hypoxia. Either extreme, the cold of a cockpit at altitude at night with a failed heating system or the green-house environment of a poorly air-conditioned and pressurized aircraft at high noon, will cause the body to expend energy in an attempt to maintain its core temperature within acceptable limits. This expended energy is just another form of increased physical activity and will decrease a pilot's tolerance to hypoxic conditions.

Self-imposed factors

Individuals vary widely in their susceptibility to oxygen deficiency and the same individual will show variances from day to day. This is primarily due to many factors influencing susceptibility to hypoxia that are under direct control of the pilot. It is the pilot's responsibility to avoid these factors as much as possible. A pilot's tolerance to oxygen deficiency will be reduced by any one or more of the following factors. The effect of combining these factors cannot be accurately assessed.

Fatigue Fatigue is both an exacerbating factor of and a symptom of hypoxia. A mentally or physically fatigued pilot will have less tolerance for hypoxia and its associated decrements in performance and perception because the fatigue will already have degraded performance, perhaps to unacceptable levels. Hypoxia will deepen the fatigue, and the cycle continues on a downward spiral of increasing fatigue and degradation of performance. Also, there is a chance that the pilot will not attribute increasing fatigue to the effects of hypoxia and will not take prompt corrective actions.

Alcohol This cellular toxin is a risk factor even after the blood-alcohol level has returned to zero. Of course, as noted previously, any alcohol in the blood or cells will hamper the ability to take up and utilize oxygen. One ounce of alcohol in the blood will raise the body's perceived altitude by 2000 feet; however, the after effects of the alcohol can be just as debilitating. The fatigue caused by the disturbance of normal sleep cycles by alcohol will diminish tolerance to hypoxia as discussed above. Additionally, the depressant effect of this drug will remain after the toxin is cleared from the blood. This will cloud the pilot's judgment and delay the recognition of the problem.

Carbon monoxide Again, it is important to remember that carbon monoxide (CO) will combine with hemoglobin in the red blood cells 200–300 times more readily than will oxygen. Once bound, it is almost impossible to rid the red blood cells of their cargo of this toxin. In fact, most carbon monoxide remains bound to the red cells until the cells die and are scavenged by the liver. Of course, the major source of carbon monoxide in the blood stream is cigarette smoke, either your own or secondhand smoke from other smokers in the vicinity.

Probably no other self-imposed risk factor is as deadly or as controllable as is the CO level in the blood. Smoking a pack of cigarettes in the 24 hours preceding a flight can saturate as much as 8–10 percent of the available hemoglobin. This will raise the body's perceived altitude by as much as 5000 feet! You can be effectively in Denver while on the ground at LAX. You will suffer the effects of hypoxia at sea level. An 8000-foot cabin of a jet cruising in the flight levels is going to be just the same as cruising unpressurized at 13,000 feet. Is it any wonder that most accidents happen in the landing phase? One wonders how much hypoxia contributes to the accident rate yet is rarely listed as a cause, especially in modern jet transports.

SYMPTOMS OF HYPOXIA

The symptoms of hypoxia are many and varied and can vary in the same individual from day to day. Many of the symptoms are very subjective and most will not indicate how hypoxic an individual has become. The following are only some of the more common symptoms exhibited by hypoxic pilots:

- Euphoria—"What, me worry?"
- Impaired judgment—make dumb mistakes, slow thinking
- Tunnel vision
- Tingling in fingers and toes
- Bluish lips and fingernails—cyanosis
- Headache
- Aggressiveness—authoritarian—confrontational with ATC, and the like
- Light-headedness, dizziness
- Fatigue—sleepy for no reason
- Slowed muscular coordination
- Diminished sense of touch and pain—your aching back feels better!
- Impairment of ability to focus eyes and accommodate to changing distances and light levels
- Muscular weakness
- Overconfidence
- Decrease in reaction time
- Greatly impaired night vision
- Depression and irritability
- Alteration in respiration—breathing slower or faster
- Sense of "tightness" in chest—short of breath

The subjectiveness of hypoxic symptoms is one of the factors that makes the recognition of this condition difficult for the one affected. Is the fatigue due to the hypoxia or the lateness of the hour or the length of the flight? Is a change in heart rate and respiration a reaction to the stress of an IFR flight with turbulence, or is it a sign of hypoxia?

Hypoxia is far more common than popularly believed or admitted. It is a trap that every pilot has fallen into at least once—whether or not the pilot is aware of the event! Each pilot will react differently to hypoxic conditions and the symptoms of the same pilot will change with changes in body chemistry and basic health. Additionally, the symptoms any one individual will experience will change with age. Military pilots are sent to the altitude chamber every three years to experience these changes in their own symptoms. Some pilots experience many mild symptoms of hypoxia, while others progress rapidly from a mild sense of fatigue to total incapacitation. You might not experience any symptoms of hypoxia at 10,000 feet today and next week you might become somewhat incapacitated at 6000 feet. There is no reliable way to predict when these variations will occur.

PREVENTION AND TREATMENT OF HYPOXIA
Prevention

There is only one surefire method of preventing hypoxia and that is to stay at the altitude to which the body is adapted and avoid all other contributors and risk factors. That would make it very difficult, if not impossible, to participate in any aviation activities. As with most hazards associated with aviation, the best policy to adopt for dealing with hypoxia can be summed up in two words: awareness and avoidance.

Awareness of hypoxia goes further than just reading the numbers. There is one gauge in the cockpit that can alert the pilot to the increasing danger of exposure to hypoxia. That gauge is the altimeter. Anytime the cabin is at an altitude higher than 5000 feet, the pilot should be aware that hypoxia will be a factor in any further inflight events. Perhaps this seems a bit extreme and premature, but once hypoxia sets in, the pilot will not have the good judgment remaining to properly assess risk. Aviation accidents rarely can be attributed to a single cause, but hypoxia is very often the unattributed event that starts the snowball effect that ends in disaster.

Avoidance of hypoxia starts with maintenance of the body in the best possible condition. A healthy pilot, not overweight, who does not smoke, drinks no more than occasionally and never the day before a flight, and who is in good condition from regular exercise will be less affected by hypoxic symptoms and will experience those symptoms at a higher altitude. This is just good preventative maintenance on the most crucial part of the flight equation, the pilot. A high degree of suspicion of any cabin altitude above 5000 feet and regular use of supplemental oxygen at those altitudes—especially at night—will go a long way toward preventing hypoxic symptoms. The watchword here is avoidance. Do not expect to wait until the symptoms of hypoxia appear. You probably will not recognize them until you are significantly compromised.

Treatment

The only treatment for hypoxia is oxygen in sufficient quantity and at sufficient pressure to adequately oxygenate the blood and supply the tissues. The pilot has two choices—either use supplemental oxygen or descend to an altitude where the atmospheric pressure is high enough that the partial pressure of oxygen is adequate for the body's needs.

Supplemental oxygen is required on all pressurized aircraft. Most systems on jet aircraft have several mask settings, two of which are 100 percent and emergency or pressure breathing. Refer to Table 3-2. Notice that the total atmospheric pressure at 40,000 feet is 140.7 mm Hg. Subtracting 46 mm Hg from this for the water vapor pressure present in the lungs leaves a total oxygen pressure of 94.7 mm Hg. As was discussed earlier, 106 mm Hg oxygen pressure is necessary to totally oxygenate the blood. Above about FL 370 pressure breathing is necessary to provide 100 percent oxygenation to the blood.

CABIN DECOMPRESSION—EXPLOSIVE AND OTHERWISE

Our fragile cocoon—those sheets of aluminum and plexiglass that give us the false sense of security as we cruise in shirtsleeves at FL 430—can fail us. It is not a thought that most pilots choose to dwell on, but the possibility is certainly there. Although structural failure like the one experienced by the Aloha Airlines 737 would certainly cause a sudden and explosive decompression, a more likely cause would be the failure of a door seal or a cracked cabin window. Explosive decompression and emergency-descent drills are practiced at least once a year by all jet pilots during recurrent training, and I am often amazed at the casual attitude shown toward these emergency situations. The actuality will not bear any resemblance to the neat and orderly drill accomplished in training.

An actual decompression will first of all be accompanied by a good deal of noise as the higher pressure air in the cabin rushes out until the cabin pressure, and consequently the cabin altitude, equals the ambient pressure outside the aircraft. This might be preceded by a loud popping sound—the sound of a champagne cork amplified 100 times. Dust and debris will be picked up and rush toward the opening where the pressurized air is rushing out of the cabin. Smaller items might be sucked outside the aircraft. A fog will form in the cabin since the warmer air in the cabin is capable of carrying more moisture than the cold air outside the aircraft. As the cabin temperature and pressure drop, the moisture will condense, forming a wet, cold fog. A significant temperature change will occur—the ambient temperature at FL 430 is –67° F. Confusion will reign. And these are only the effects on the atmosphere inside the craft—what of the effects on the humans?

First of all, there is a distinction between rapid and explosive decompressions. Any decompression that takes place in less than one-half second is considered to be an explosive decompression. This type of decompression occurs primarily in smaller-bodied aircraft and, thankfully, is not common since it can be rapidly fatal. Human lungs usually require about two-tenths of a second to release their air. Any decompression happening in less than this short amount of time can result in rapid lung decompression and rupture or severe damage. The only emergency procedure available for dealing with this type of decompression is to immediately get on oxygen and get the aircraft to a lower altitude as quickly as possible.

A rapid decompression occurs in more than one-half second but less than about 10 seconds. This is the type of decompression experienced in larger-bodied aircraft and is the more common of the two. There is not the high potential for lung damage in this type of decompression, however, the noise, confusion, debris, and fogging will all be present in varying degrees of severity. Rapid donning of oxygen and descent is still necessary, but other emergency measures might also be available to deal with the situation and lessen the impact of the decompression.[2]

Subtle decompression is also a danger in pressurized aircraft. A gradual loss of cabin pressure (or, more commonly, improper setting of the cabin altitude controls) and a slowly rising cabin altitude might not be recognized by the crew in time to deal with the emergency effectively. In all jet aircraft certified to transport category standards, FAR 25.841 requires "Warning indication at the pilot or flight engineer station to indicate when the safe or preset pressure differential and cabin pressure altitude limits are exceeded. . . . And an aural or visual signal (in addition to a cabin-altitude indicating means) meets the warning requirement for cabin pressure altitude limits if it warns the flight crew when the cabin pressure altitude exceeds 10,000 feet." In addition, these aircraft are required to have means to limit the cabin altitude to no more than 15,000 feet in the case of a failure of components in the pressurization system. This cabin altitude limitation, however, does not refer to structural failures, but only to the components of the pressurization system itself. (See chapter 5 for a discussion of the pressurization system.)

Effects of decompression on the body

Of course, the most serious effect of decompression is the resultant hypoxia and more or less rapid loss of effective consciousness. The most noticeable immediate effect of rapid cabin decompression will be the sudden rush of air from the lungs. As noted above, this can be a near-fatal experience in explosive decompressions in small aircraft. It will, however, happen. I've had pilots tell me they would be able to hold their breath and prevent this effect. That won't happen. First of all, the surprise of the event will override any defensive measure you might have thought you could put into place. Secondly, the rapid change in pressure differential will make it all but impossible to hold your breath. (Remember, we can be talking about a pressure differential of 8.8 PSI. Let's put these pressure differentials into perspective. A differential of 8.8 PSI is 1267 pounds per square foot. The pressure differential between the bottom and the top of the wing of a popular corporate jet is less than ⅓ PSI, and that supports the weight of the aircraft!)

Trapped gases

Recall that Boyle's law states that the volume of any gas is directly proportional to the pressure exerted on that gas. In other words, as the pressure drops, the gas expands. Any gas trapped in the human body will expand with a drop in the pressure surrounding the body. This can occur in various places in the body causing varying degrees of discomfort or pain.

Ear blocks The trapped gas disorder that almost everyone who has ever flown is familiar with is one affecting the ears, actually the middle-ear areas. Usually, trapped gas in this area is a problem on descents, but discomfort and pain can be present during rapid decompressions. The eustachian tube connecting the middle ear to the nasal passages normally acts to equalize the pressure between the outer ear and the middle ear. (These two areas are separated by the eardrum. See Fig. 3-6.) Any pressure differential between these two areas will cause the eardrum to bulge, and the compromised flexibility of the eardrum will affect hearing.

Swelling of the eustachian tube, common with "head colds," will prevent normal equalization of pressure between the outer and middle ear. Because the nasal end of the eustachian tube acts as somewhat of a one-way valve allowing air to pass out of the middle ear, this problem is more common on descent than on ascent

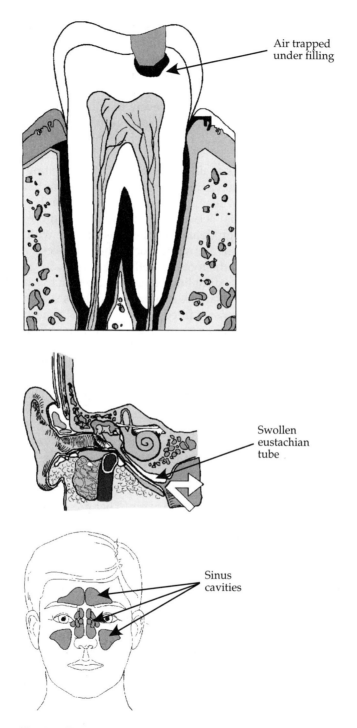

Air trapped
under filling

Swollen
eustachian
tube

Sinus
cavities

Fig. 3-6. Trapped-gas disorders can occur beneath dental fill-
ings, in sinus cavities, and in the middle ear.

or rapid decompressions, but can happen, especially in the presence of severe swelling of the mucus membranes lining these passages. Because the greatest change in air density occurs between sea level and 5000 feet, these problems will be more severe with rapid pressure changes at these altitudes than in the rarer air of the flight levels.

Sinus block Sinus block can be more severe than blockage of the eustachian tube leading to the middle ear because the passages between the sinuses and the nasal cavity are much smaller than the eustachian tube. This problem can be equally severe during ascent or descent. With a rapid decompression and any inflammation of the sinus passages at all, severe, almost incapacitating pain might be felt. Some have described this pain as feeling as though a nail was being driven into the cheekbone. The only relief available is descent to a higher ambient air pressure to relieve the pressure differential between the sinus cavity and the ambient air.

Dental problems Although dental problems associated with rapid decompression are not as common as ear and sinus blocks, they can occur. Any abscess or infection in the gums or around the roots of teeth will cause pain upon ascent and a rapid decompression can cause disabling pain. Only a return to a higher ambient pressure will bring relief, and dental care should be sought as soon as possible.

Teeth that have been improperly filled cause problems at altitude or during rapid decompressions. The higher pressure under the filling will cause excruciating pain, and in rare instances can cause the tooth to explode. An exploding tooth would be distracting, not to mention the pain associated with the failure.

Intestinal problems Intestinal problems associated with rapid decompression range from merely embarrassing to totally incapacitating. There is normally about one quart of free air in the intestinal tract. This is air that is swallowed and gases that are produced by digestive processes and fermentation. Diet variations can increase or decrease this average volume. This air, too, will obey Boyle's law and increase in volume with the decreasing ambient air pressure. That quart of air, at sea level, will expand to more than nine quarts at 43,000 feet. (The expansion is greater than that shown in Fig. 3-2. Moist air, as is found in all body cavities containing air, expands at a greater rate than dry air.) That greatly expanded volume of air can cause severe intestinal cramping and pain, or it can be simply a discomfort. Much depends on the fatigue level, apprehension, and general physical condition. The expanding gases will attempt to escape through both ends of the gastrointestinal tract.

It is well known that many foods produce more than the normal share of intestinal gases, beans being the most familiar example. Other foods such as cabbage, cucumbers, radishes, onions, raw apples, and some varieties of melons can cause increased intestinal gases and should probably be avoided 24 hours before flight. (Avoiding everything that could be a potential hazard or discomfort in rare emergency situations would make a very dull life at times and is probably an unrealistic goal.) Carbonated beverages and gum chewing (because of the increased air swallowed by gum chewers) should also be avoided during flight.[3] (Water is probably the best drink for pilots wanting to stave off the inevitable dehydration associated with high-altitude flight. Coffee and other caffeinated beverages should be avoided because of their diuretic effect.)

EVOLVED-GAS DISORDERS

The bends is a common term usually associated with scuba and other kinds of divers. It is, however, only one of several kinds of evolved-gas disorders. These are medical conditions, also called decompression sickness or DCS, associated with the release of dissolved gases into the body. If you will recall Henry's law, you will remember that the amount of gas dissolved in a solution is directly proportional to the pressure of the gas over the solution. This is the effect observed when the top is removed from a carbonated beverage. The release of pressure over the solution allows the carbon dioxide to bubble out of solution.

The human body contains a great deal of nitrogen (an inert gas comprising 78 percent of the atmosphere) dissolved in the blood and other body tissues. This nitrogen has been introduced into solution at sea-level, or near to sea level, atmospheric pressure. Any pressure above this will allow the nitrogen to come out of solution. Flight in an unpressurized aircraft or a pressurized aircraft with a cabin altitude above sea level will allow the nitrogen and other inert gases to come out of solution in small bubbles.[4]

Effects of evolved gases on the body

Circulatory System The formation of gas bubbles in the circulatory system is perhaps the potentially most serious of the evolved-gas disorders. The gas bubbles, properly called *aeroembolisms*, first block the smallest capillaries, but as the bubbles become larger, larger and larger vessels become involved. Blockage of blood vessels in the heart, lungs, or brain can be rapidly disabling, even fatal. Damage to body organs will vary with extent and duration of blockage.

Chest and lungs Gas bubbles in the chest (thorax) area and vessels of the lungs cause a condition and symptoms known as *the chokes*. These symptoms begin with a burning pain in the central area of the chest that progresses to a stabbing pain and is worsened by deep breathing. An almost uncontrollable cough will be unproductive and continued time at altitude will bring on feelings of suffocation and typical signs of cyanosis.[5] The lips, ear lobes, and fingernails turn blue and tingle. At this point, if an immediate descent to a higher ambient pressure is not made, collapse and death are possible.

Muscles and bones Nitrogen bubbles forming in the fluid that fills the joint spaces, especially the larger joints of the shoulders, elbows, and knees cause varying amounts of pain ranging from an annoying ache to severe and incapacitating pain. This is the same affliction suffered by divers ascending too rapidly from deep dives. Popularly called *the bends*, this condition will become steadily worse until equilibrium is reached between the pressure of gases dissolved in the blood and the ambient pressure. Again, the only cure for this affliction is descent into an area of higher ambient pressure. Subsequent exposures to lower ambient pressure than the body is acclimated for will tend to cause recurring pain in the same joint spaces first affected.

Nervous system Evolved-gas disorders affecting the nervous system range from the mildly annoying tingling, itching, or cold and warm feelings caused by bubbles of nitrogen formed around the nerve tracts in the skin to the life-threatening air embolism in the brain. Peripheral nerve involvement rarely causes permanent damage; however, the symptoms are almost identical to those presented by hypoxia and hyperventilation.

Central nervous system (brain and spinal cord) involvement are much more serious. Early symptoms of nitrogen bubble formation in these areas are usually visual disturbances such as the appearance of flashing or flickering lights, headaches, and confusion. More severe and potentially life-threatening symptoms can include partial or total body paralysis, loss of hearing or speech, and unconsciousness. The appearance of any of these symptoms heralds a medical emergency and descent to a higher ambient pressure and immediate medical intervention are necessary to preclude permanent disability or death.

Factors affecting severity

Several factors affect the potential severity of decompression sickness. These factors can be divided into two broad categories: those pertaining to the environment and those pertaining to the individual.

Environmental factors The environmental factors affecting the severity of decompression sickness are the altitude of exposure, the time spent at that altitude, and the rate of ascent. As in all the effects suffered by the human body due to altitude, the higher the altitude, the worse the effects of exposure. Decompression sicknesses are not usually experienced below 18,000 feet, and, as with hypoxic symptoms, the severity increases rapidly above the middle 20,000s.

DCS symptoms will increase proportionately to the time spent at altitude. As long as the body is exposed to ambient pressures below those of acclimatization, nitrogen will continue to bubble out of solution and collect in the body, just as the carbon dioxide will continue to come out of solution in a glass of soda, collect on the side of the glass, and diffuse into the atmosphere until the soda is "flat."

Another factor affecting the severity and onset of DCS symptoms is the number of exposures. Although one would not anticipate having several rapid cabin decompressions in a short period of time, several trips to high altitude in a non-pressurized aircraft will increase the likelihood of decompression sickness. Additionally, repeated exposures of affected organs will make subsequent episodes of DCS more severe and more likely to occur.

Personal factors Personal factors affecting the onset and severity of DCS symptoms include the age of the individual exposed and the basic state of health enjoyed. Older persons tend to be more susceptible to DCS at lower altitudes and with shorter exposure times than younger individuals, other health factors being equal. Of greater importance, however, is the basic health of the individual. Obese people will store much more nitrogen in fat tissue than their leaner colleagues, as much as five to six times more. Obviously, the chance of developing severe DCS symptoms is greater in these individuals.

Any injury to joints or muscles will increase the likelihood and the severity of decompression symptoms by making the release of the dissolved nitrogen easier. Additionally, recent strenuous exercise will lower the threshold for DCS impairment. Again, we are caught between a rock and a hard place. That morning run will certainly improve your cardiovascular fitness and help stave off both hypoxia and DCS from a aerobic standpoint, but will make the occurrence of DCS more likely because of the stress put on the musculoskeletal tissues. Any heavy exertion just prior to or immediately after a high altitude flight is not wise.

The self-imposed risk factor that is easiest to avoid is scuba diving. Because diving exposes the body to pressures above that found on the surface, more than the normal amount of nitrogen will be dissolved in the body fluid and tissues. Flying, even without a cabin decompression, immediately after diving can bring on symptoms of DCS at altitudes as low as 8000 feet (the normal altitude of a transport-category aircraft at maximum cabin differential at altitude). A rapid decompression after scuba diving can be rapidly life threatening.

THE ALTITUDE CHAMBER

The FAA's physiological training, also known as the altitude chamber ride, is a required and recurring experience for military aircrew members and is offered to civilian pilots at various military bases and at the FAA's Mike Monroney Aeronautical Center in Oklahoma City. Although completion of an altitude-chamber ride is no longer required[6] by the FAA for any certificate or rating for civilian pilots, this training is invaluable, and it is strongly recommended that all pilots, not just those flying in jets or other pressurized equipment, have at least one chamber ride as early in their training as possible. No amount of reading, attendance at lectures and ground schools, or listening to the experiences of others will prepare you to detect your own symptoms of hypoxia. You must have the actual experience of your own symptoms before you reach the disturbance stage and cannot, or will not, take corrective action.

All pilots, whether flying jets or ultralights, should be encouraged to experience the altitude chamber as early in their flying career as possible for the following reasons:

- To gain knowledge about the causes and effects of hypoxia.
- To experience the basic symptoms of hypoxia and decompression in a controlled environment.
- To experience and recognize personal symptoms of hypoxia before the condition becomes disabling. Because each person responds differently to the lack of oxygen, the chamber ride is the only safe method to gain insights into individual unique symptoms.
- To experience rapid decompression under controlled conditions.
- To gain respect for the insidious effects of hypoxia and their reality in every flight environment, especially the high-altitude environment.

CONCLUSION

In conclusion, awareness and respect for hypoxia, its causes, and effects are central to the health and safety of every pilot. The atmosphere is comprised of several regions, each being less compatible to life than the one below it. These regions are easily charted and recognized. The gases in the atmosphere unfailingly obey the laws of physics. The tissues in the human body unfailingly react to diminished pressures of oxygen and total atmospheric pressure. The conscious brain of the individual pilot—that organ of his or her body most demanding of and dependent upon a rich supply of oxygen at sufficient pressure—is hard to convince of the seriousness of the problem.

Many of the symptoms of jet lag (fatigue, irritability, muscle aches and headaches) are attributable to hypoxia, yet never presented as such. If the passengers are hypoxic, does it not stand to reason that the crew is also? All pilots have had more than one encounter with hypoxia. Few realize and fewer still will admit it. Perhaps it is not coincidence that many aircraft accidents happen during the landing phase when the crew's ability has been eroded by lack of sufficient oxygen.

Try this experiment. The next time you have an opportunity to fly at a cabin altitude of 8000 feet or above for several hours, take along an oxygen supply. After at least two hours at altitude, take careful note of your performance and how you feel, then put on the oxygen. After several minutes breathing with the oxygen supply, again assess your performance and state of physical well-being. You will note a dramatic difference. This experiment is especially eye-opening (pun intended) at night when the true extent of the degradation of vision by lack of oxygen can be appreciated. If this demonstration does not make a believer out of you, I'm not sure what would.

Endnotes

1. The typical cabin altitude of a corporate jet cruising at FL 450 is 8000 feet.

2. These will be dictated by the cause of the decompression and will be found in the aircraft flight manual.

3. Do not avoid drinking fluids during flight. Water or fruit juices should be consumed in ample quantity to avoid the dehydration that is inevitable in the arid atmosphere of a pressurized cabin. Coffee should be avoided because of its diuretic effects.

4. Fat tissue will dissolve five to six times as much nitrogen as can be dissolved in the blood. This is another reason to maintain fitness and eliminate extra body fat; it can become a true hazard to life and health in the event of a sudden cabin decompression.

5. Cyanosis is a bluish discoloration of the skin caused by excessive venous, or oxygen-poor, blood. Hemoglobin is red when carrying oxygen, bluish when not. Carboxyhemoglobin, which is hemoglobin with carbon monoxide rather than oxygen bound to it, is very bright red. Victims of carbon monoxide poisoning almost always have rosy cheeks and lips.

6. The altitude chamber was at one time required for the single-pilot exemption to fly Cessna Citation II and S/II aircraft at takeoff weights greater than 12,500 pounds.

4

Jet engines

EVEN THE NONAVIATION PUBLIC TODAY HAS A BASIC UNDERSTANDING of jet engines. It is commonly known that these engines propel aircraft by sucking air in the front and expelling hot exhaust gases from the rear. Unfortunately, this is also the understanding of jet engines mastered by most pilots of piston-powered aircraft. Although the operation of these engines is not really much more complicated than this rudimentary explanation, an understanding and appreciation for the theory and operating characteristics of these efficient powerplants will do much to explain the operation of jet aircraft in general.

PHYSICS

Having some understanding of the applicable principles of physics to gain an understanding of jet propulsion principles and the operating characteristics of jet engines is necessary. The physics described here is not an exhaustive treatment of the principles involved but is rather to present the basic ideas necessary for an understanding of the relationships of the gases and the machinery within a gas-turbine engine. The mass flow of gases referred to is simply atmospheric air compressed and accelerated in the turbine engine to create work at the turbine wheel and thrust. The thrust is created either by a fan driven by a turbine or by the reaction to the gases flowing from the exhaust, or a combination of the two.

Some of the most important physical properties that apply to gas turbine, or jet engines are weight, mass, density, temperature, and pressure. These properties are used to define and discuss ideas such as force, work, acceleration, thrust, and others that measure valuable turbine performance factors. Let us briefly discuss each of these properties and ideas and the units used to describe them before we launch into their interactions in jet engines.

Weight

Weight is a unidirectional force that is a measure of the attraction between a body and the Earth. It is a measure of gravitational attraction and is usually expressed in pounds of force or kilograms of force.[1] The force called weight always acts in the same direction, toward the center of the Earth. Weight is known as a *vector quantity* since it has both size and direction. Figure 4-1 illustrates the idea of weight.

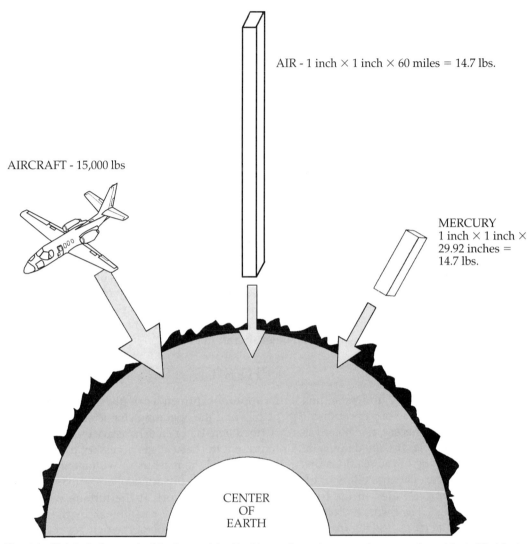

AIR - 1 inch × 1 inch × 60 miles = 14.7 lbs.

AIRCRAFT - 15,000 lbs

MERCURY
1 inch × 1 inch ×
29.92 inches =
14.7 lbs.

CENTER
OF
EARTH

Fig. 4-1. Weight is the result of the force of the Earth's gravity acting upon the mass of an object. All objects attract each other with a gravitational force, but the attraction that the aircraft has for the Earth is minuscule compared to the attraction that the Earth has for the aircraft. All weight is directed toward the center of the Earth.

Mass

Mass is how much material a body contains. This idea is often confused with weight, but mass is independent of the force of gravity. A one-kilogram mass on the surface of the Earth will "weigh" one kilogram (actually, 9.8 N). A one-kilogram mass on the surface of the Moon will weigh only ⅙ kilogram or 1.63 N (9.8 ÷ 6) since the gravitational force on the Moon is only ⅙ that of the Earth. The mass of the object, however, will remain at one kilogram no matter where in the universe it is found.

To be correct, one defines mass as a measure of inertia or that tendency of an object either to remain at rest or maintain a state of uniform motion in a straight line unless acted upon by an outside force. Mass is an inherent property of a body and is independent of the body's surroundings and the means used to measure mass. It will take just as much force to displace a 100-kilogram anvil in the "weightlessness" of space as it would on the surface of the Earth.

Mass is also often confused with the volume of an object. As Fig. 4-2 illustrates, a one-kilogram mass of feathers and a one-kilogram mass of lead occupy vastly different volumes, but both exhibit the same inertia, or resistance to change in movement. The differences in the two volumes can be explained by the relative densities of the two materials.

Density

Density, illustrated in Fig. 4-3, means the mass of a substance per unit volume. The units of density in the scientific world are kilograms per cubic meter or grams per

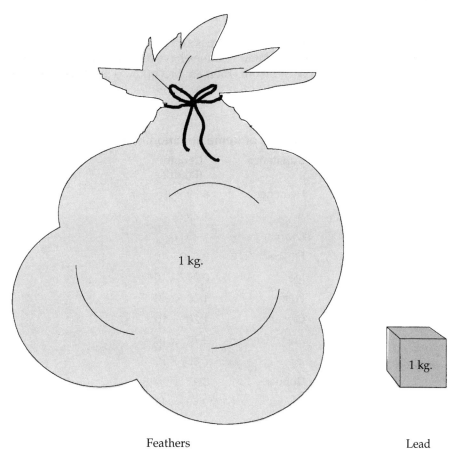

Feathers   Lead

Fig. 4-2. A bag of feathers and a block of lead: Both have the same mass.

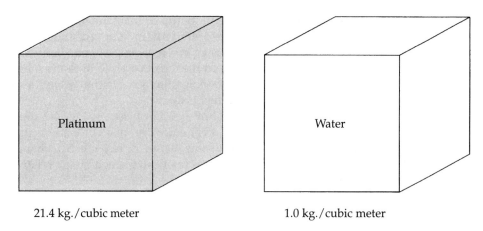

Platinum	Water
21.4 kg./cubic meter	1.0 kg./cubic meter

Fig. 4-3. Platinum is far denser than water. One cubic meter of platinum has a mass of 21.4 kg., while 1 cubic meter of water has a mass of 1 kg.

cubic centimeter, but any unit that relates mass and volume can be used. Two containers of the same material with equal volumes have different weights if the contents are packed more closely in one than in the other. The density of some common substances is listed in Table 4-1.

Table 4-1. Density of some common materials

Substance	Density (kg/m³)
Hydrogen	8.99×10^{-2}
Air	1.29
Oxygen	1.43
Ethyl alcohol	0.806×10^3
Ice	0.917×10^3
Water	1.00×10^3
Glycerin	1.26×10^3
Aluminum	2.70×10^3
Iron	7.86×10^3
Copper	8.92×10^3
Silver	10.50×10^3
Lead	11.30×10^3
Gold	19.30×10^3
Platinum	21.40×10^3

The compressor of a gas-turbine engine uses this principle and packs more molecules of air into a given space. This increases the density and the weight of the volume of air and therefore the thrust produced by the engine. For example, on a standard day the weight of uncompressed air at sea level is 0.076475 lbs./cubic foot. In an engine with a compressor ratio of 20-to-1 the weight of one cubic foot of air will be 0.076475 × 20 or 1.5295 lbs.

Temperature

Temperature is described in Fig. 4-4 as the molecular energy of motion due to heat.[2] At low temperatures, molecular motion is low, and at higher temperatures, molecular motion increases. Higher temperatures are a problem in the compressor of a gas-turbine engine since it requires more work (in terms of compressor speed

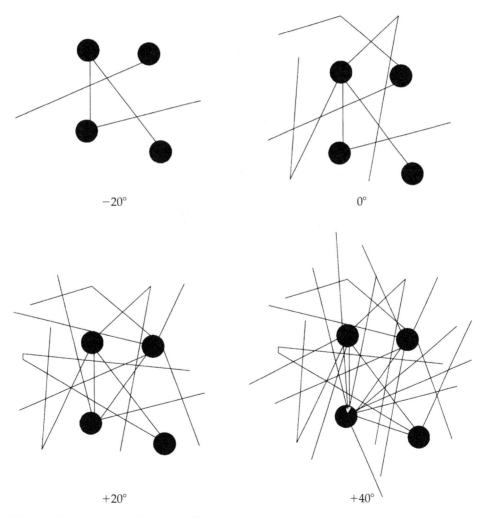

Fig. 4-4. Temperature is the energy of motion.

and consequently fuel consumption) to increase the density of air if the temperature of the air is increased.

Pressure

Pressure is an omnidirectional force produced by the motion of molecules.[3] Air molecules are rebounding off the inner walls of the container with such great rapidity that an essentially constant pressure is exerted over all of the inner surfaces. Pressure is defined as force per area on which it acts ($P = F/A$). Pressure of gases will be equal in all directions, confined or unconfined, and is usually measured in pounds per square inch (PSI or lbs/in^2). Figure 4-5 illustrates the idea of pressure. Changes in pressure are central to most aviation discussions.

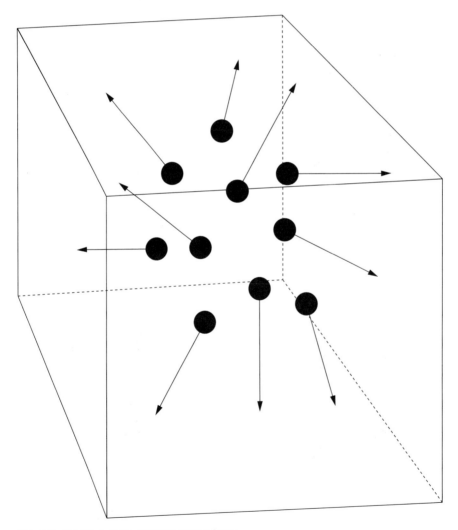

Fig. 4-5. Pressure is an omnidirectional force.

Force

Force is defined in physics as any influence that can cause an object to be accelerated and is measured in Newtons in the scientific community. Force is the product of the pressure applied and the area to which it is applied ($F = PA$). Force and pressure are often confused, but reference to Fig. 4-6 will show that while both 1-lb. bricks exert the same force on the surface of the table, the pressure exerted is different. The force needed to accelerate an object is the product of the mass of that object in kilograms and the acceleration in meters per second per second according to the formula $F = ma$. Force as measured in turbine engines is normally expressed in pounds although in the world of physicists, the Newton is the preferred unit of force as previously noted.

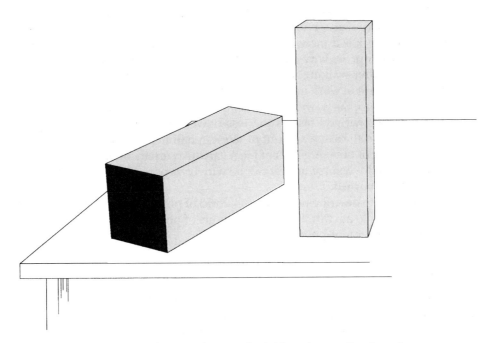

Fig. 4-6. Both blocks exert the same force on the table as long as they have the same mass. The block standing on end, however, exerts greater pressure on the table.

Work

Work is done when a force acting on a body causes it to move through a distance.[4] Notice that movement is an integral part of this definition. No matter how hard you push against a brick wall, and no matter how tired you become, you will not have done any work (on the wall) if it does not move. (You will, however, have done work on yourself. Muscle fibers will have moved and energy will have been expended.) Additionally, since work is only measured along the direction in which the force is applied, carrying a bucket full of water over 10 miles does no work on the bucket of water since the force applied to the bucket is perpendicular to the direction of movement of the bucket. (Again, the muscles in your arm will tell you a totally different story!)

Work is described as useful motion with weight times distance components when performed in the vertical direction and force times distance components when performed in the horizontal direction according to the formula $W = F \times D$. In physics, this is expressed as *joules*, one joule being the force of one Newton applied over the distance of one meter. When applied to aircraft engines, work is usually expressed in units of foot-pounds or ft. lbs., with one ft. lb. being one pound of force exerted over the distance of one foot.

Power

Power is the rate of doing work. In the definition of work, no mention is made of time. Applying the dimension of time to work gives us the concept of power. The same amount of work is done when carrying a load up a flight of stairs whether we walk up or run up. So, why are we more tired after running up the stairs in a few seconds than after walking upstairs in a few minutes? Running up the stairs does the same amount of work in less time and therefore takes more power. An aircraft engine with twice the power of another does not necessarily produce twice as much thrust or go twice as fast as the less-powerful engine. Twice the power means that it will do the same amount of work in half the time or twice the work in the same amount of time. A gallon of Jet-A can do a certain amount of work, but the power produced when we burn it can be widely varying amounts, depending upon how fast it is burned.

The units used to describe power in the world of physics are joules per second (J/s). When applied to aircraft, however, the units of power are usually expressed in units of ft. lbs./s, or foot-pounds per second.

Velocity

Velocity uses the same units as speed, meters per second in physics, and miles per hour or feet per second for the rest of us. Although velocity and speed are often confused, and in fact the terms are often used interchangeably in conversation, they are not only different quantities, they are different types of quantity. Speed is defined as distance over time and is a scalar[5] quantity. Velocity is a vector quantity and must be defined with magnitude and direction. If we wish to describe the velocity of an aircraft, we must specify its speed (say 400 knots) and the direction in which the aircraft is moving (say north). Displacement and acceleration are vector quantities we will be using in our discussion of jet engines.

Acceleration

In physics, acceleration is defined as the change in velocity over time. Distance traveled is not an issue, only increase or decrease of velocity with time. An important acceleration present in the physical world is the acceleration due to the Earth's gravity. This is a force of attraction between the Earth and all other objects. The magnitude of this acceleration near the surface of the Earth (which for our purposes will be the case) is 32 ft./s².

As with the concept of pressure, discussion of accelerations is central to almost any aviation discussion. Acceleration of air allows the wings to support the weight of the aircraft and the engine to move the aircraft forward.

ENERGIES—POTENTIAL AND KINETIC

Most discussions of energy are quite vague. We speak of the energy necessary to get a job done, the energy necessary to run a marathon, the "energy crisis," and various other uses of the term. Rarely, however, is the actual definition of energy contemplated.

The combination of energy and matter make up the universe. The concept of matter is easy to grasp. Matter is substance. It is "stuff." It can be seen and touched. Matter has mass and occupies space. Energy, on the other hand, is more abstract. We cannot see, smell, or touch most forms of energy. Although energy seems to be something we are familiar with, it is difficult to define because it is not just a "thing" but rather a thing and a process.

People, places, and things all have energy, but we only observe energy in a roundabout manner. Energy has sometimes been defined as the ability to do work. It is somewhat like pornography, to be defined only by example. We see energy only when it is happening, only when it is being transformed. Energy reaches the Earth in the form of electromagnetic waves from the Sun and we feel this energy as heat. The food we eat contains stored energy and we transform this energy by the process of digestion. Aircraft fuels contain energy that is transformed in the process of combustion.

One key concept about energy has led to one of the greatest generalizations in physics. The Law of Conservation of Energy states: "Energy cannot be created or destroyed: it may be transformed from one form into another, but the total amount of energy never changes."

That is heavy stuff, but in any closed system—be that a swinging pendulum or a total galaxy—the total amount of energy will remain unchanged. That energy will be a constantly changing state, but the total will remain the same.

Potential energy Potential energy is the energy an object possesses due to its position. A stretched or compressed spring possesses potential energy, or the potential to do work. Water stored behind a dam and the elevated ram of a pile driver possess potential energy. When the string on a bow is drawn, energy is stored in the bow. The energy stored in fuels is potential energy for it is energy due to position on a molecular level.

Kinetic energy Kinetic energy is the energy of motion. When an object is set into motion, by whatever means, work is done on that object to change its energy of motion. When an object is in motion, it is capable of doing work by virtue of that motion. The kinetic energy of an object depends upon its mass and its speed according to the formula $KE = \frac{1}{2}mv^2$, where KE is kinetic energy, m is mass, and v is speed.

When an object is in motion, work has been done upon that object to put it into motion. The kinetic energy of an object is equal to the work done upon it to bring it from rest to that speed, or the work a moving object can do in being brought to rest.[6] Notice that the energy of an object can be varied directly by varying that object's mass. Varying the object's speed, however, will vary the energy by the square of the increase. Doubling the speed of a moving vehicle will require four times as much energy to stop that vehicle, a concept of physics well known to crash investigators and highway patrolmen!

Bernoulli's theorem

It's time to revisit Mr. Bernoulli (1700–1782), and this time our discussion will be more closely aligned with the studies of this famous Swiss physicist. As you can see from his lifespan, Daniel Bernoulli knew nothing about airplane wings, the item associated with his name by most pilots. Bernoulli's most famous work, *Hydrodynamica*, published in 1738, was a study of the equilibrium, pressure, and velocity of fluids. This work also attempted the first explanation of the behavior of gases with changing temperature and pressure and was the beginning of the kinetic theory of gases.

Simply stated, Bernoulli discovered that air acts as an incompressible fluid[7] when flowing at subsonic speeds and that as the velocity of a fluid flow increases, its pressure decreases. Also, the sum of the static and dynamic pressures of a fluid in a duct remains constant. Pressure can also be changed in the gas in a turbine engine by adding or removing heat, changing the number of molecules present, or by changing the volume in which the gas is contained.

When a gas is supplied at a constant rate of flow through a duct, the sum of the potential and the kinetic energy of the gas will remain constant. When the dynamic (ram) pressure of the gas increases, the static pressure will decrease. Figure 4-7 illustrates Bernoulli's theorem. We assume a constant flow rate in pounds per second through the duct at points A, B, and C. The gauges in the stream at points A, B, and C will measure total pressure. The gauges above points A, B, and C measure only the static pressure of the liquid: the potential energy. Subtracting the static pressure from the total pressure will give the ram pressure, or kinetic energy of the fluid, at any point. (Remember, the total energy of any system remains constant according to the Law of Conservation of Energy.)

NEWTON'S LAWS AND JET ENGINES
Newton's first law

Newton's first law of motion states that "a body at rest will tend to remain at rest and a body in motion will tend to remain in unaccelerated motion unless acted upon by an outside force." In other words, a force is needed to accelerate a mass. All aircraft engines, be they piston or turbine, accelerate varying masses of air. A propeller provides thrust to an aircraft by a relatively small acceleration of a large parcel of air. The turbojet imparts a much greater acceleration to a smaller quantity of air. The turbofan, being somewhat of a hybrid of the two, combines the small ac-

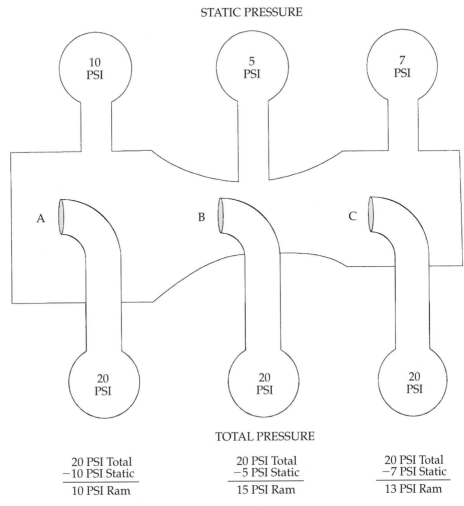

STATIC PRESSURE

| 10 PSI | 5 PSI | 7 PSI |

A B C

| 20 PSI | 20 PSI | 20 PSI |

TOTAL PRESSURE

20 PSI Total	20 PSI Total	20 PSI Total
−10 PSI Static	−5 PSI Static	−7 PSI Static
10 PSI Ram	15 PSI Ram	13 PSI Ram

Fig. 4-7. Total pressure is the sum of the static pressure and the ram pressure, or pressure of motion. Assuming a constant flow through the pipe, the static pressure will decrease at the constriction in the pipe and increase when the pipe widens.

celeration of a large quantity of air of the propeller in its fan stage with the large acceleration of a small quantity of air by its jet core.

Newton's second law

In his second law, Newton states that a force is equal to the product of a mass and the acceleration given to that mass, or $F = ma$. Now, units of mass are difficult to work with mathematically, but the assumption can be made that mass and weight are interchangeable when an object is near the surface of the Earth. (Physicists will argue with this; however, the results of the calculations made in this manner are accurate enough for our purposes here.) If a one-pound object exerts a one-pound force on a scale due to the attraction (acceleration) due to gravity, then:

$$F = m \times a$$

where: $a = 32 \text{ ft./s}^2$

$1 \text{ lb} = m \times 32 \text{ ft./s}^2$

$F = 1 \text{ lb.}$

$m = 1 \text{ lb}/32 \text{ ft/s}^2$

$m = F/a, \text{ or } m = W/g$

In other words, when dividing weight by the acceleration due to gravity, we end up with mass units that can be used in our calculations. (The attraction, or acceleration, due to gravity acts on all matter near the surface of the Earth, including air passing through the engine.)

We can now expand our formula for force, or thrust, into an even more useful form in which we can substitute readily measurable quantities:

$$F = W/g \times V_2 - V_1/t$$

where: W = weight of air

V_2 = velocity of air leaving engine

V_1 = velocity of air entering engine

t = time

g = gravity constant

F = lbs. of force or thrust

Newton's third law

Newton's third law of motion is commonly stated "for every action there is an equal and opposite reaction," but Newton actually said that "if two bodies interact, the force exerted on body 1 by body 2 is equal in magnitude and opposite in direction to the force exerted on body 2 by body 1." It is important to note that this law of physics relates to *reaction pairs*; both action and reaction do not take place on the same object.

In a turbine engine, the engine pushes on the air and the air in turn pushes on the engine. Because the engine is fixed to the airplane, the whole combination moves through the air. (This is really no different than the way propeller engines move the airplane. All current aircraft engines move air; only the volumes and degree of acceleration differ.)

TYPES OF JET ENGINES

There are four common types of jet engines: rocket jet, ramjet, pulse-jet, and turbine-type jet. The turbine-type jet is the type found on civilian aircraft and is the type we will be concerned with. The turbine family contains four types of engine: turboprop, turboshaft, turbofan, and turbojet. We will limit our discussion to the turbofan and turbojet types.

Turbojet engines

A turbojet engine is a reaction engine that produces thrust by taking in air, compressing it, adding fuel, and combusting it, creating a flow of hot gases. This flow of gases is used to turn the turbine wheel, and the remaining gases in the tail pipe accelerate into the atmosphere and create the reaction we refer to as thrust. Figure 4-8 shows a simple diagram of a turbojet engine and illustrates its main sections: intake, compression, combustion, and exhaust.

Although most subsonic aircraft in use today employ turbofan engines, the pure turbojet is still in use on the supersonic Concorde and some military craft. Its relative inefficiency, when compared to the turbofan, and the higher noise levels associated with its operation have caused the turbojet to fall into disfavor for commercial use.

Turbofan engines

A turbofan engine (Fig. 4-9) is simply a turbojet engine with a fixed-pitch propeller in the front (It is usually in the front because rear-fan engines are much less efficient and therefore rarely used.). The fan acts as a propeller and gives a relatively small rearward acceleration to a large volume of air. The turbojet engine core gives a relatively large amount of acceleration to a small package of air. This combination engine preserves the low-speed, low-altitude efficiency (when contrasted to turbojet engines) of propeller engines with the high-speed, high-altitude efficiency

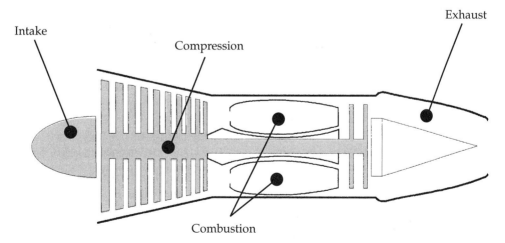

Fig. 4-8. A typical turbojet engine showing the intake, compression, combustion, and exhaust sections.

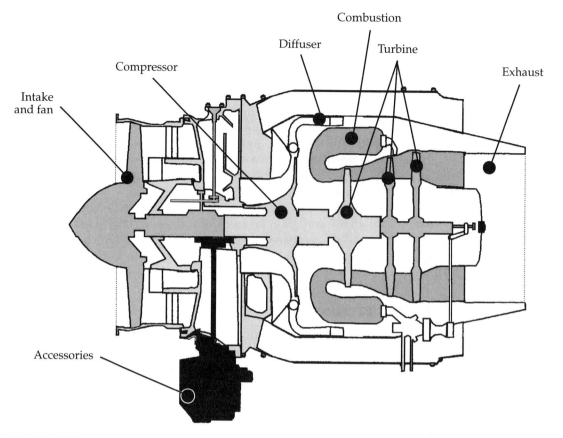

Fig. 4-9. A Pratt & Whitney JT15D-1A turbofan engine used on the Cessna Citation I aircraft.

Combustion

Diffuser

Turbine

Exhaust

Compressor

Intake
and fan

Accessories

of jets. That is, with its ducted design, the turbofan has turbojet-type cruise speed and altitude capability and yet retains the short field takeoff capability of the pro-peller-driven craft.

Turbofans are as much as 30 to 40 percent more fuel efficient than turbojet en-gines for the following reasons:

- The turbofan wastes a lower amount of kinetic energy from its fan exhaust. This is due to the fact that the average speed of the combination fan exhaust and turbine core exhaust is closer to the speed of the aircraft.

- There is a lower amount of kinetic energy left in the atmosphere after the aircraft has passed. Consider the following hypothetical example:

An aircraft engine expels 10 mass units of air from its exhaust at a velocity of 1000 ft./s. Kinetic energy left in the atmosphere will follow the formula

$$KE = \frac{1}{2} mv^2$$

$$KE = \frac{1}{2} \times 10 \times 1000^2$$
$$KE = 5{,}000{,}000 \text{ ft. lbs. wasted energy.}$$

If the same engine expels the same mass at double the velocity, the following will result:

$$KE = \frac{1}{2} mv^2$$
$$KE = \frac{1}{2} \times 10 \times 2000^2$$
$$KE = 20{,}000{,}000 \text{ ft. lbs. wasted energy}$$

Now, if the engine is redesigned to expel double the mass of air at the same velocity as the original engine, we find the following:

$$KE = \frac{1}{2} mv^2$$
$$KE = \frac{1}{2} \times 20 \times 1000^2$$
$$KE = 10{,}000{,}000 \text{ ft. lbs. wasted energy}$$

As can be seen from the above examples, the greatest propulsive efficiency is gained by giving the smallest amount of acceleration to the largest parcel of air. A turbofan engine achieves this goal while maintaining high-altitude efficiency. This discussion should also lead the reader to the conclusion that high exhaust velocity and low aircraft speed are a highly inefficient operating condition.

Bypass ratio

Turbofan engines are classified according to their bypass ratio as low-, medium-, and high-bypass turbofans. Bypass ratio compares the amount of air that bypasses the jet core of the engine to that routed to the core.

In a low-bypass turbofan, the fan and the compressor sections receive and utilize approximately the same amounts of air. This, then, would be indicated as a bypass ratio of 1-to-1. Military turbofans usually have a bypass ratio of 1-to-1 or less due to the narrow profile requirements of supersonic flight.

Medium- or intermediate-bypass turbofans have bypass ratios in the range of 2- or 3-to-1. It should be noted, with all turbofan engines, that the ratio of thrust provided by the fan to the ratio of thrust provided by the jet core will be approximately the same as the bypass ratio.

High-bypass turbofans boast the lowest fuel consumption of the various turbofan configurations and have bypass ratios in the area of 5-to-1. The thrust produced by the fans in these engines is 75 percent to 85 percent of the total thrust output of the engine. Very few high-bypass fans are fully ducted due to the weight penalty involved in a wide diameter fully ducted design (Fig. 4-10). The high-bypass turbofan design has gained wide popularity and use for medium to large airliners because of greater propulsive and thermal efficiency and consequent greater fuel economy.

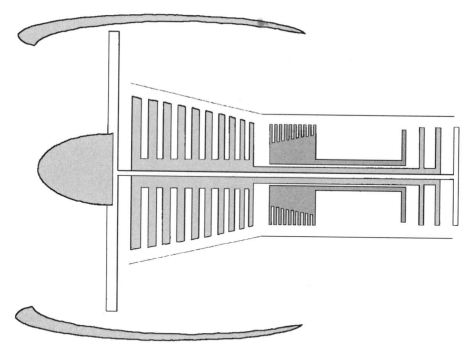

Fig. 4-10. A high-bypass-ratio turbofan engine is rarely fully ducted because of the weight penalty associated with fully enclosing the large bypass duct.

JET PROPULSION ADVANTAGES

It is widely believed that turbine aircraft consume large amounts of fuel while reciprocating powerplants seem to be more fuel conscious. If the simple comparison of relative gallons or pounds of fuel per hour is used, this misleading conclusion might seem to be true. If the comparison is made, however, between ton/miles of payload per unit of fuel consumed, the jet would be seen to be more efficient.

If the Boeing 747 could somehow be fitted with piston engines—although it would take 23 of the largest piston engines ever produced to equal the 230,000 lbs. of static trust provided by the four JT9D turbofans—the fuel consumption of the combination would be much higher than that of the jet-powered craft. This increased fuel consumption would be largely the result of the increased weight and drag of the piston engines; however, propeller efficiency losses would also contribute to the penalty.

Some of the major advantages of jet propulsion are:

- Lower weight and drag allowing higher speeds and larger payloads.

- Virtual absence of engine vibration caused by reciprocating parts allows airframes to be lighter.

- Increased efficiency with altitude obviating the need for complex supercharging systems.

- Production is cheaper and speedier. A turbine powerplant has approximately one-fourth the parts of a reciprocating powerplant.

- Increased reliability and lowered maintenance costs due to fewer parts, constant combustion, and absence of reciprocating parts.

- Engine efficiency goes up with rising aircraft speed due to the influence of ram pressure increasing the engine mass airflow and exhaust velocity.

JET PROPULSION THEORY

SUCK

SQUEEZE

BANG

BLOW

Jet propulsion theory in four words, and yet, in truth, it is not really much more complicated than that. In fact, even four-stroke piston engines operate under much this same theory. Air must enter the engine intake (suck). The air must then be compressed (squeeze), either by the upstroke of a piston or the compressor section of a jet engine. The fuel is introduced and the fuel-air mixture is ignited (bang) in the cylinder of a piston engine or the combustion chamber of a jet, and the hot expanding gases are exhausted (blow). Because a jet engine has separate areas for these four functions, the functions can all occur simultaneously and provide constant rather than intermittent power (and subsequent vibrations) provided by a piston engine.

Jet engine terminology and parts

Figure 4-11 illustrates a simple turbofan engine. This engine, the JT15D used on the Cessna Citation and Beechjet aircraft, is characterized as a medium-bypass, twin-spool, axial-flow, turbofan engine. The bypass ratio of 3.3-to-1 puts this engine on the verge of being a high-bypass engine and, in fact, 75 percent of the thrust of this engine is produced by the fan. The term *twin-spool* describes the two concentric shafts of the engine. The N_1 shaft, the longer and consequently central of the two, connects the low-pressure turbine to the N_1 fan. The outer and shorter shaft connects the high-pressure turbine to the N_2 compressor. The engine is called an *axial-flow*, even though the compressor is of the centrifugal type since the mass airflow through the engine is along the axis of rotation of the engine. (Larger engines use multistage axial compressors instead of or in addition to a centrifugal compressor, but the terminology remains the same.) Last of all, the engine is a turbofan engine, realizing a portion of its thrust from a ducted fan stage rotated by exhaust gases.

Intake and fan—SUCK

The inlet duct on most commercial jet engines is of the fixed geometry, divergent type. A divergent duct has an increasing diameter from front to back and allows

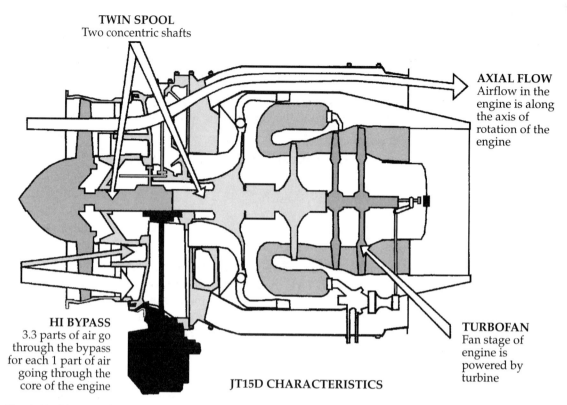

TWIN SPOOL
Two concentric shafts

AXIAL FLOW
Airflow in the engine is along the axis of rotation of the engine

HI BYPASS
3.3 parts of air go through the bypass for each 1 part of air going through the core of the engine

JT15D CHARACTERISTICS

TURBOFAN
Fan stage of engine is powered by turbine

Fig. 4-11. Diagram of the P&W JT15D turbofan engine showing the characteristics used to describe this type of engine.

the incoming air mass to diffuse at the entrance to the duct and progress at a constant pressure to the fan stage. This diffusion in the inlet increases the static pressure of the incoming air and reduces the dynamic pressure or velocity. Although the inlet ducting is usually considered an airframe part rather than an engine part, the inlet is usually referred to as *engine stage number one.*

The engine inlet's appearance belies its importance. The compressor section requires a uniform supply of air if it is to enjoy stall-free performance. It has been shown that even a small discontinuity or turbulence in the airflow supplied to the compressor will cause significant losses in propulsive efficiency and many engine performance problems. The inlet duct must also create as little drag as possible.

Inlet pressures add significantly to the mass airflow through the engine as the aircraft reaches cruising speeds. With the aircraft stationary, the pressure within the inlet will be slightly below ambient due to the demands of the compressor. (Remember, suck!) As the aircraft accelerates during takeoff, ram pressure recovery takes place. At approximately Mach 0.1 or 0.2, the pressure at the compressor inlet will have returned, or recovered, to ambient. As the aircraft accelerates in flight, the pressure, or ram recovery, at the inlet will continue to increase and will effectively increase the compression ratio of the compressor without an additional expenditure of fuel. The net result will be an increase in thrust without an increase in fuel flow. The faster a jet flies, the more efficient the engine becomes. The ram compression ratio of a typical business jet cruising at Mach 0.8 is in the neighborhood of 1.5-to-1.

The fan stage of a turbofan engine also provides a compression ratio of about 1.5-to-1. The majority of the mass airflow from the fan passes over the outer portion of the fan blades, through the bypass duct, and back to the atmosphere. The majority of the thrust from the engine comes from this air. The remainder of the air passes over the inner portion of the fan blades, through a set of stators, and into the compressor stage of the engine. The speed of the N_1 fan in a twin-spool engine will be seen to increase with altitude as the atmospheric density decreases and consequently the drag on the fan is reduced.

Compressor Section—SQUEEZE

The compressor section of the engine operates to supply the combustor with sufficient quantities of air (in terms of mass). The primary purpose of the compressor is to receive mass airflow from the inlet and to increase its static pressure before discharging it to the diffuser, and, then, to the combustor at the correct temperature, pressure, and velocity. Secondarily, compressors supply engine bleed air to cool hot-section parts, to pressurize bearing seals, and to supply heated air for inlet and fuel system anti-icing. Bleed air is also extracted for those items not associated with the operation of the engine such as aircraft pressurization, air conditioning, and other functions, collectively known as *customer-service bleed air*, requiring clean, pressurized air.

Compressors operate by accelerating air to increase its kinetic energy followed by diffusion to translate the acquired kinetic energy into potential energy. The efficiency of a compressor is based upon the principle of maximum compression with minimum temperature rise, and modern engines show efficiencies in the 85 to 90 percent range.

Centrifugal-flow compressors The centrifugal-flow compressor is the oldest and simplest of the two compressor designs. This compressor design, which resembles a washing machine agitator (Fig. 4-12), receives its airflow at the center of the impeller and accelerates the air outward by centrifugal reaction to its rotational speed. The air is then routed to a diffuser section and, in accordance with Bernoulli's principle, as the air spreads out, its speed is reduced and its static pressure is increased. The compressor manifold then delivers the turbulence-free air to the combustion chamber.

Centrifugal-flow compressors, although modern designs are capable of compression ratios of 10-to-1, are not practical in more than two stages due to energy loss in the airflow when making the turns from one impeller to the next and the added weight penalty in multiple stages. Their use, therefore, is limited to the smaller engines found on small corporate jets, rotorcraft, turboprops, and auxiliary power units. Centrifugal compressors are commonly used in conjunction with axial-flow compressors and are positioned as the final compressor stage. All larger engines today are of the axial-flow compressor type.

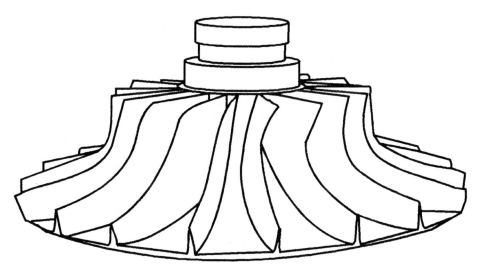

Fig. 4-12. A typical centrifugal-flow compressor.

Tip speeds in centrifugal compressors can reach approximately Mach 1.3; however, the radial airflow remains subsonic. The pressure within the compressor case is sufficient to prevent airflow separation at low supersonic rotor speeds and additionally causes a high energy transfer to the airflow. Although the large frontal area, and consequent drag, is a disadvantage of the centrifugal compressor, the advantages of this design are as follows:

- Low weight and consequently low starting power requirements.
- Simplicity of design and manufacture compared to axial compressors and consequently lower cost.
- High pressure rise per stage, up to a 10-to-1 ratio.
- Good efficiency over a wide range of rotational speed from idle power to full power.

Axial-flow compressors Axial-flow compressors are so named because the direction of the mass airflow through the compressor is parallel to the rotational axis of the compressor (Fig. 4-13). They are normally of several stages, each stage comprised of two components, the rotor and the stator. Although there are three main types of axial flow compressors—the single-spool, dual- or twin-spool, and triple-spool—the twin-spool design is by far the most popular in current production engines. Limited use of the triple-spool design is just becoming evident as of this writing.

Single-spool engines have one rotating mass comprised of the compressor, shaft, and turbine rotating together as a single unit. Most fan engines today are of the twin-spool type and have two concentric, or coaxial, shafts. The longer shaft connects the N_1, or low-pressure, compressor to the N_1, or low-pressure, turbine wheel(s). The inner and shorter of the two shafts connects the N_2 high-pressure compressor to the high-pressure turbine wheel(s). An additional variation of the

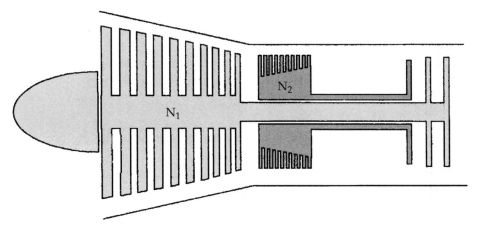

Fig. 4-13. A typical axial-flow compressor.

twin-spool engine uses a reduction gear box to convert turbine speeds into torque to turn the fan. The Garrett TFE-731 is this type of engine. Triple-spool engines have compressors numbered N_1, N_2, and N_3, and termed, in order, low-pressure, intermediate-pressure, and high-pressure compressors.

The advantages of dual- and triple-spool axial compressors can be seen in the operational flexibility they provide for the engine in terms of quick acceleration, higher compression ratios, and improved stall characteristics. For any given power setting, the high-pressure compressor speed is held fairly constant by the fuel control unit. Because a fairly constant energy level is available at the turbine, the low-pressure compressor(s) will change speed in response to changing conditions at the engine inlet. In other words, the N_1 compressor tries to supply the N_2 compressor with a constant air pressure for a given power setting. This flexibility is not possible with single-spool designs.

As the temperature of the air increases, the air's molecular motion increases. In order to provide a constant supply of air to the high-pressure compressor, the low-pressure compressor would either have to change the angle of its blades (not possible) or vary its speed, which it does. Also, as altitude increases and air density decreases, the speed of the N_1 compressor increases as a result of the reduction of drag from the rarified air. This supercharges the N_2 system.

Each stage of an axial-flow compressor is comprised of one rotor and one stator. The *stators*, or stationary vanes, are placed to the rear of the rotor blades and act as diffusers to change the kinetic energy of the air into potential energy. The stators also direct the airflow to the next stage of compression at the desired angle. Axial-flow compressors in use today normally have from 10 to 18 stages of compression.

As pressure builds in the rear stages of the axial compressor, air velocity tends to slow according to Bernoulli's principle. This is undesirable because a gas-turbine engine operates on the principle of velocity change in airflow. In order to stabilize the velocity of the airflow undergoing the pressure changes, the shape of the compressor gas path converges as it moves aft, reducing its volume to approximately 25 percent that of the inlet-flow area. This tapered shape prevents the diffusion and consequent velocity drop in the airflow.

There are many advantages and, in contrast, many disadvantages to the use of axial compressors. Chief among these are the following advantages:

- Smaller frontal area and consequent lower drag than centrifugal compressors.

- Higher pressures and higher peak efficiencies attainable by addition of compressor stages.

- High compressor pressure ratios created by straight-through design.

These advantages are somewhat offset by the following disadvantages:

- Good compression only in the cruise through takeoff power settings.

- Many stages needed due to low compression rise per stage.

- Difficulty and high cost of manufacture.

- High weight due to multiple stages needed.

- High starting power requirements to accelerate large mass of compressors.

Combination compressors In order to take advantage of the best features of both types of compressors, many smaller engines use a combination-type compressor, which is a multistage axial compressor followed by a centrifugal compressor as its last stage. This arrangement takes advantage of the low frontal area and drag of the axial portion and couples it with the good compression ratio over a wide operating range of the centrifugal portion. This combination compressor is especially well suited to engines with a reverse-flow annular combustion chamber such as those found on the Pratt & Whitney PT-6 series of turboprop engines and the P&W JT15D series of turbofans.

Compressor stall Compressors stall for the very same reason that wings stall—the critical angle of attack has been exceeded. The angle of attack of a compressor blade is the result of the combination of inlet velocity and compressor speed. When the critical angle of attack is exceeded, airflow through the compressor slows down, stagnates, or even reverses direction. The stall condition can range from the transient and hardly noticeable to the attention-getting "hung stall," which manifests itself by loud backfire-type noises and severe fluctuation in cockpit gauges for engine RPM, EGT, and fuel flow.

Severe compressor stalls can cause sufficient back pressure from reverse airflow through the engine that pressure on the rear of compressor blades causes them to bend forward and contact the stator vanes of the preceding stage. This can lead to a progressive series of material failures and total disintegration of the rotor system and catastrophic engine failure. Although compressor stalls can result from a variety of conditions, the most common causes are:

- Turbulent or disrupted airflow through the engine inlet.

- Engine operation above (or below) designed limiting revolutions per minute.

- Excessive fuel flow caused by abrupt engine acceleration—especially at higher altitudes.
- Excessively low fuel flow caused by abrupt deceleration.
- Damaged turbine components causing loss of power to the compressor and low compression.
- Contaminated or damaged compressor.

Some situations pilots might find themselves in that can cause one or a combination of the above conditions include acceleration for takeoff down a runway with a rapidly shifting and gusting wind of high velocity. An abrupt shift to a strong crosswind coupled with engine acceleration can cause a stalled condition in the compressor. Abrupt pitch-up and power application for a balked landing or missed approach can also provoke a stalled compressor(s). Practice maneuvers such as recoveries from approaches to stalls can cause abrupt pitching maneuvers and compressor stalls. Any foreign object damage (FOD)[8] increases the likelihood of compressor-stall conditions.

The pilot will be made aware of the stalled condition of the compressor by the noise and the fluctuations of engine gauges, primarily the exhaust-gas temperature and turbine RPM gauges. Usually, stabilizing the pitch attitude of the aircraft and retarding the power to allow inlet-air velocity and revolutions per minute to return to their proper relationship will correct the condition. Severe compressor stalls are more likely the result of faulty fuel-control units or damage to the compressor or turbine and cannot be so easily corrected from the cockpit.

Compressor-diffuser section The section of the engine between the compressor and the combustion chamber is known as the compressor-diffuser section. This section provides additional space for the air coming off the compressor to spread out and increase its static pressure. The compressor-diffuser is usually a diverging duct and is known as the point of highest pressure in a gas-turbine engine.

The term "point of highest pressure" needs some interpretation. If the total pressure of the air at the diffuser inlet is 200 PSI, the total pressure of the exit will also be 200 PSI. Remember, however, that the total pressure (P_t) is comprised of the static pressure and the dynamic, or ram pressure of the air. Because the diffuser section is designed to slow the mass airflow and increase the static pressure (P_s), the ram pressure of the air will decrease, and the total pressure will remain the same. It is this increase in static pressure that is referred to when the diffuser exit is called the point of highest pressure in the engine.[9]

Combustion Section—BANG

The combustion section, or burner can, consists of an outer casing, an inner liner, a fuel-injection system, and a starting ignition system. Although the combustor would seem to be the simplest of the components of a gas-turbine engine, the art of combustor design has sometimes been referred to as "black art." In other words, engineers are not always sure why one combustor design works and another doesn't when installed on the same engine. Obtaining good combustor performance and service life still takes the bulk of engine research and developmental time and money, much the same as in the early days of gas-turbine engine development.

Combustor operation

The job of the combustion section is to take the air delivered by the compressor and add heat energy to the flowing gases by the addition of fuel and a starting source. This addition of heat energy causes expansion and acceleration of gases into the turbine section. When fuel heat is added, the resulting volume of gas is increased, and with the flow area remaining the same, the gases must accelerate.

To function properly, a combustion chamber must provide for the proper mixing of the air and fuel and also cool the hot combustion products to a temperature that the turbine section and, in fact, the combustion chamber itself, can withstand. Because turbine fuel, unless propelled into the ground at high speeds and thoroughly vaporized, is notoriously hard to ignite and has slow flame propagation times, this job is much more difficult than it might sound.

As we have already seen, 75 to 85 percent of the air entering the inlet of a modern turbofan is routed through the bypass ducting. Of the remaining 15 to 25 percent of the air routed to the jet core, only about 25 percent is used in the actual combustion process; thus, for every pound of air entering the engine inlet, about one ounce is ultimately mixed with fuel and ignited. The rest of the air is used to cool the burner can itself and the turbine section of the engine.

The air actually used in the combustion process is termed *primary air* and it is introduced into the burner can via two routes. About one-half of the primary air enters the burner can axially through swirl vanes in the area around the fuel nozzles. The rest of the primary air is introduced into the burner can radially through small holes in the first third of the can, and both axial and radial primary air are used in the combustion process. The velocity of the primary air in the flame zone immediately in front of the fuel nozzles must be carefully controlled. Although the velocity of the secondary air through the burner can might reach the speed of several hundred feet per second, the primary air is almost stagnant in order to provide the required mixing time for air and fuel. Due to the slow flame-propagation time of jet fuels, if primary air velocity were too high, it would blow the flame out of the engine and cause a *"flame-out."* Although flame-out is an uncommon event in modern engines, it still occurs for a variety of reasons. Turbulent weather, heavy precipitation, high-speed maneuvers, and rapid power reduction at high altitude are some of the more typical reasons for flame-out. As can be seen, two of these conditions, rapid power reductions and high-speed maneuvers, are under the direct control of the pilot. The other two are either avoidable or controllable to some extent.

Combustion occurs in the first third of the combustor liner. In the remaining two-thirds of combustor length, the products of combustion are mixed with approximately half of the secondary air to provide a temperature at the turbine inlet that is consistent with long service life and equal distribution of heat. The remaining secondary air is used as a cooling shroud around the inside and outside of the combustion liner, cooling the surface of the liner and centering the flame to prevent it from contacting the metal surfaces. Any disturbance of this secondary cooling air can result in holes burned in the burner can and sharply decreased component life.

Combustor efficiency in modern gas-turbine engines is in the 99 to 100 percent range. In other words, the combustor extracts heat equal to 99 to 100 percent of the potential heat energy actually contained in the fuel. Not a bad record for a "black art!"

Types of combustors

Multi-can (through flow) This type of combustor is older in design and is not commonly used today. It consists of multiple burner cans, each with its own outer case and inner liner interconnected by flame propagation tubes and arranged around the engine. Each can is, in effect, a separate combustor unit, all of which discharge into a common area just in front of the turbine inlet. The flame propagation tubes allow the flame, ignited in the two cans with the ignition source, to progress to the other individual units. The Rolls-Royce Dart engine uses this type of burner-can technology.

Can-annular (through flow) This design utilizes one outer case with several combustion liners. Each liner has its own fuel nozzles, and all liners are connected by flame propagation tubes. The liners take in air at the front and discharge it at the rear. Two bottom liners contain the igniter plugs. This design, found on many commercial aircraft powered by Pratt & Whitney engines, is designed for ease of on-wing maintenance.

Annular combustor (through flow) This design takes in air at the front and discharges it at the rear and consists of one outer housing and one liner. The annular burner can is somewhat donut-shaped and surrounds the engine rear of the diffuser section. This design is used in all sizes of engines and is considered to be the most efficient of the combustor designs. It is the most efficient in terms of thermal energy versus weight, and the smaller surface area of the liner interior requires less cooling air.

Annular combustor (reverse flow) This design is characteristic of smaller Pratt & Whitney engines, the Avco-Lycoming T-53/55, and several other low-mass airflow engines found on corporate aircraft. The reverse-flow combustor has the same function as the through-flow combustors, but the air flows from the diffuser over the outside of the burner can and enters the combustion chamber in the rear. The expanded and accelerated exhaust gases are directed to a deflector that turns them 180 degrees to exit the rear of the engine. The main advantage of this type of burner can is the shorter engine length allowed by positioning the turbine wheels inside the area of the combustor rather than in tandem with it as in the through-flow engines. This arrangement also allows for preheating of the compressor discharge air that makes up for the loss in efficiency caused by the turns made by the gases during combustion. (See Fig. 4-14.)

Turbine and exhaust—BLOW

After the exhaust gases exit the combustion chamber, they pass through two or more turbine stages on their way to the engine exhaust. The turbine stage transforms a portion of the kinetic and heat energy in the exhaust gases to mechanical work, enabling the turbine to drive the compressor and the accessory section of the engine. The compressor adds energy to air by increasing its pressure. The turbine extracts

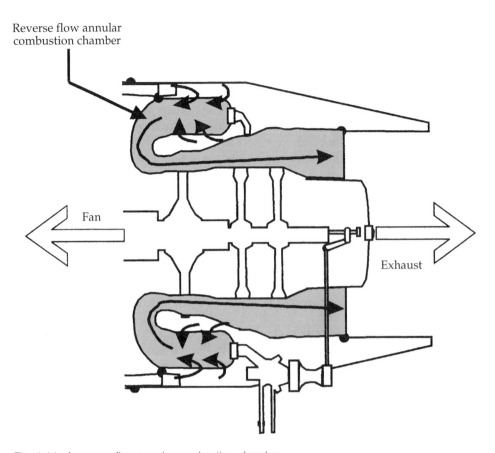

Reverse flow annular
combustion chamber

Fan

Exhaust

Fig. 4-14. A reverse-flow annular-combustion chamber.

this energy by reducing the pressure of the flowing exhaust gases. The first stage turbine wheel(s) is the N_2, or high-pressure, turbine, which powers the high-pressure compressor.

The mass of the airflow, obviously, does not change in the transfer of energy from the combustion chamber to the turbine wheels, but the energy of the flow is reduced. The pressure of the exhaust gases is converted to velocity at the nozzles formed by the turbine blades and stator vanes, and these gases are vectored tangentially to the rotor. This slows the gas flow axially (front to rear), but adds shaft power to the rotor system. In other words, tangential velocity is a loss of kinetic energy to the engine, but is a gain of energy to the turbine. The exhaust gases pass next through the low-pressure (N_1) turbine stage(s) that power the low-pressure compressor and fan stages. These turbine stages work to convert kinetic energy into mechanical energy in the same manner as the high-pressure turbine.

The turbine blades and stator vanes are cooled by compressor discharge air, which is also known as *bleed air*. There are several cooling arrangements; two main schemes in use are internal-airflow cooling and surface-film cooling. In the internal-airflow cooling arrangement, bleed air flows through hollow blades and vanes and carries the heat away after convection cooling. The surface film cooling also uses bleed air routed to the interior of hollow blades and vanes, but this air then flows from small exit ports in the leading and/or trailing edges to form a heat barrier on the surfaces. This cooling air, from both arrangements, is then discharged into the engine airflow at the cooling location. This cooling of the turbine blades and stator vanes allows these components to function in a thermal environment 600 to 800°F above the temperature limits for the metals and alloys used for blade and vane construction. Maximum turbine inlet temperatures (TIT) of approximately 3000°F are common in engines of modern design due to this cooling of blades and vanes.

After extraction of energy by the turbine stages, the exhaust gases are discharged through the exhaust cone, tail cone, and tail pipe. The exhaust cone is sometimes referred to as the *exhaust collector* and collects the exhaust gases discharged from the turbine discharge. The tail cone acts as a diffuser within the exhaust cone and serves to reduce the turbulence downstream of the turbine wheel. The exhaust struts that support the tail cone act to return the airflow to an axial direction.

The tail pipe is an airframe part and is a convergent duct that causes the exhaust gases to accelerate to the design speed necessary to produce the required thrust. The convergent-shaped duct can accelerate the exhaust gases to Mach 1, at which time the nozzle is said to be *"choked."*

SUPPORTING SYSTEMS AND ENGINE ACCESSORIES

Various systems, or subsystems if you will, support the engine in its main task of accelerating air. The fuel system, ignition system, engine starting system, lubrication, and bleed-air systems are all integral to this job. Engine anti-ice, fire detection, and fire protection systems support and protect the engine.

Engine accessories such as fuel pumps, fuel control units, oil pumps, hydraulic pumps, starter-generators, and tach generators are mounted on the accessory gear pads of the engine. These pads, located in the accessory gearbox, are powered by a tower shaft from the N_2 compressor turning a bevel gear. Various gearing arrangements and ratios, dependent upon engine manufacturer and design, can be found, but all operate in essentially the same manner.

Engine fuel system

Although there will be variations in engine fuel systems found on the many types of aircraft powered by gas-turbine engines, the main operating principles remain the same across the various platforms.[10] Fuel must be delivered from the aircraft storage and delivery system to the engine at a pressure and in a quantity sufficient to support combustion over a wide range of engine operating speeds.

Jet fuels are liquid hydrocarbons that are similar to kerosene, some of which are blended with gasoline. Jet fuels are not color coded as are reciprocating engine fuels but are a natural straw color and have a characteristically "greasy" feeling. Turbine fuels require additives to retard bacterial growth and icing. Although all aviation fuels contain some dissolved water, jet fuels are hygroscopic, meaning that they readily attract and absorb moisture from the air. This dissolved water does not pose an icing risk, but it does make the hydrocarbon fuel an excellent breeding ground for all sorts of microorganisms that can clog fuel filters and other system components.

Water that is not dissolved but only held in suspension in the fuel is referred to as *entrained* water, and it is this entrained water that poses an icing risk. Beside the risk of icing at higher altitudes, large amounts of entrained water (over 30 parts per million) can cause engine performance loss or engine flameout. Most turbine fuels marketed today are premixed with anti-icing and antimicrobial agents. PRIST is a popular brand of hand-servicing agent for those fuels that are not pre-mixed. The engine manufacturer will specify the type and amount of additive to be mixed with the fuel.

Fuel pump

The engine-driven fuel pump receives fuel from the aircraft fuel storage and de-livery system and provides fuel pressure to the fuel control unit and subsequently to the engine fuel nozzles. This pump is designed to deliver a quantity of high-pressure fuel to the fuel control unit in excess of the engine's needs. Extra fuel is then routed back to the engine-driven pump. Fuel must be supplied to the com-bustion chamber at a high pressure in order to ensure the proper spray pattern and atomization against the pressure found in the burner can, and this requires a pres-sure of from 500–1500 PSI at takeoff power settings.

Most main fuel pumps are spur-gear, positive-displacement pumps with sin-gle or dual elements. These pumps, very similar to the gear-type oil pumps, de-liver a fixed quantity of fuel per revolution. Because engine-driven fuel pumps are self-lubricated on the fuel they pump, a means of providing constant head pres-sure to the pump must be provided to prevent cavitation and damage. This is usu-ally accomplished through the use of *motive flow*, or "jet," pumps with electrically driven pumps as backups. The fuel to operate the motive flow is obtained from the extra fuel provided from the engine-driven pump to the fuel control unit. These pumps, usually found in the fuel storage tanks, route high-pressure fuel through a venturi creating a lower pressure to pick up larger quantities of fuel to pressurize the fuel system. (See Fig. 4-15.)

Fuel control unit

This engine-driven unit, which can be of the hydro-mechanical, electro-hydrome-chanical, hydro-pneumatic or electronic type, has one function in the engine fuel system—it is basically a very expensive density-altitude computer. A fuel control unit meters fuel on the basis of weight and provides a supply of fuel to the burner can that will ensure a 15-to-1 air to fuel ratio[11] by weight for the proper combus-tion. This is the chemically correct, or *stoichiometric*, mixture for complete combus-tion of the fuel. Fuel is metered on the basis of weight rather than volume because the volume of a given weight of fuel is variable based upon temperature; however, fuel will provide a constant level of energy per unit weight.[12]

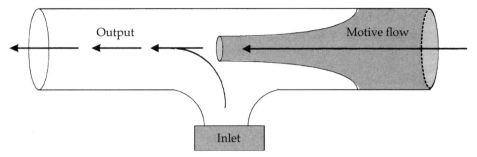

Fig. 4-15. A motive-flow, or jet-type, pump. This pump takes a small volume of high-pressure fuel from the engine-driven fuel pump and through venturi action creates a low-pressure area and picks up a larger volume of fuel at a lower pressure.

The fuel control unit typically receives several signals to be used in its control of the fuel-air ratio. Among the most common are engine speed, inlet-air pressure, compressor discharge pressure, inlet-air temperature, burner-can pressure, and power-lever position. As you can see, if you know the temperature and pressure of the inlet air and the speed of the compressor, you will be able to calculate the weight of the mass airflow coming from the compressor. Of course, if you are given compressor discharge pressure or burner-can pressure, the calculation is done for you. Power-lever position tells the fuel control unit what power setting the pilot is requesting.

The fuel control unit, now knowing the weight of the air being provided by the compressor, can compute the proper amount of fuel to meter to the fuel nozzles to maintain the 15-to-1 air to fuel ratio in the burner can. Automatic compensation is made for the normal-service bleed-air requirements of the engine and aircraft. Depending upon the sophistication of the fuel control unit and the electronic inputs it receives, it might also be able to compensate for the extraction of bleed air for engine anti-ice and ground air-conditioning requirements. If the fuel control unit is not able to make these compensations, they must be made manually by the pilot; the flight manual and checklists will so advise. To ignore manual power adjustments for these additional bleeds will risk, at the very least, a reduction in power from that expected and possibly an overtemperature condition in the burner can and at the turbine inlets. Remember, a jet engine always has far in excess of the air required for fuel combustion; however, the engine requires this air for cooling purposes. If bleed air is extracted without an adjustment to the fuel scheduled to the engine, the amount of cooling air will be reduced.

Because it takes increased compressor output and consequently increased compressor RPM to convince the fuel control unit to add more fuel to the fuel-air mixture and because it takes more fuel to increase the compressor RPM, you can readily see why turbojet engines have a spool-up lag with power application of 5–10 seconds. Engines with electronic fuel control units suffer less lag than hydromechanical types, although there will always be some lag between a request for and delivery of power. Because typical gas-turbine engine compressor speeds range in the neighborhood of 34,000 RPM, it can be seen that inertia alone will cause a lag in spool-up time.

Fuel heaters

Not all engines use fuel heaters, but of those that do, many use the engine fuel-oil cooler as a fuel heater (Fig. 4-16). This heat exchanger routes engine oil through a radiator composed of many small tubes carrying fuel to cool the engine lubricant and consequently also heat the fuel.[13] Other systems in use for fuel heating are electric elements and bleed-air heaters. The fuel heater, of whatever type, is usually positioned between the boost pumps in the aircraft fuel storage system and the main engine-driven pump to prevent icing of the fuel-pump inlet-filter screen.

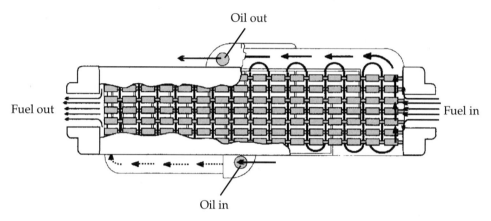

Fig. 4-16. Heat exchanger-type oil cooler. This oil cooler uses engine fuel as a means of transferring heat from the oil.

Use of fuel heat is usually restricted to 1 minute in every 30 or some similar time period due to the potential for fuel vaporization and vapor lock or damage to the fuel control unit from excessive heat. Fuel heat is not used for takeoff, approach, or go-around because of the possibility of flameout from vaporization during these crucial flight regimes.

Fuel filters

Gas-turbine engine fuel systems typically have at least two levels of filtration. A low-pressure, coarse mesh filter or screen is generally installed between the storage tank and the engine-driven fuel pump, and a much finer filter or screen is between the engine-driven pump and the fuel control unit. The fuel control unit is the point in the system most vulnerable to contaminants in the fuel due to its many small orifices, passageways, and minute tolerances. Most fuel-filtering systems incorporate a bypass around the first screen or filter and a cockpit warning to alert the pilot of actual or impending fuel-filter bypass. The action to be taken upon the illumination of a fuel-filter-bypass cockpit warning light will be dictated by the individual aircraft flight manual and can range from simply noting the condition for ground maintenance to selecting fuel heat and/or fuel transfers.

Fuel nozzles

The most common type of fuel nozzle found on modern gas-turbine engines is the *concentric duplex* nozzle. This nozzle provides fuel from two manifolds, the primary and secondary, to ensure the proper amount of fuel flow for various engine speeds and altitude requirements. Typically, the primary flow is used for starting and lower power settings with the secondary being added for acceleration and higher power settings. These nozzles are spaced evenly around the combustion chamber, 12 being the most common number found. (See Fig. 4-17.)

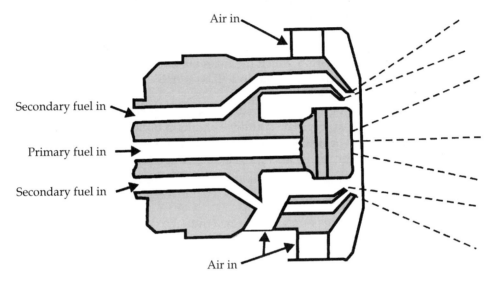

Fig. 4-17. Duplex fuel nozzle.

The fuel nozzles must atomize or vaporize the fuel and allow it to mix with the primary combustion air for proper burning. This job must be accomplished against the internal pressure of the burner can, typically in excess of 150 PSI. Any disturbance of the atomization process or spray pattern will cause inefficient burning of the fuel and consequent losses of power and potential damage to the burner-can liner.

Ignition system

Typical turbojet ignition systems are comprised of two parts: the exciter box and the igniter plugs. The exciter box receives power from the aircraft electrical supply and transforms this relatively low-voltage power (24–29 volt dc is typical on business jets or 100–120 volt ac is found on large commercial engines) to a higher voltage of 2000 volts for the low-tension systems up to 20,000 volts for high-tension systems. The igniter plugs (there are usually two installed in each engine in the lower part of the burner can) look and function similar to piston-engine spark plugs and provide between 60 and 100 sparks per minute.

A gas-turbine engine, once ignited, is self-sustaining unless an interruption in either the air or fuel supply occurs; consequently, these engines do not require a constant source of ignition as does a piston engine. Ignition will typically only be activated on a gas-turbine engine, after the start sequence, when there is a risk of flameout, or when the occurrence of a flameout would put the aircraft in jeopardy. This would include flight in turbulence or heavy precipitation, potential icing conditions, any maneuvering that might involve abrupt attitude changes, such as stall recognition and recovery maneuvers during pilot certification flights and for take-off and landing. Because the forward speed of the aircraft will cause the engine to continue to rotate (much the same as a windmilling propeller), a source of ignition will facilitate an engine relight once the air and/or fuel flow is reestablished. Engine manufacturers provide an inflight start envelope (Fig. 4-18) showing the altitude and airspeed ranges within which the mass airflow through the engine and the output of the engine-driven fuel pump will be suitable to effect a relight after a flameout or engine shutdown.

Ignition systems are designed with redundancy. All systems have two igniter plugs, and many have two separate sources of ignition excitement. Only one igniter is necessary to start or sustain a gas-turbine engine, and in most installations, the pilot will not be aware of the loss of service of one igniter plug or system.[14]

Engine starting systems

Systems used for starting gas-turbine engines must be capable of spinning the compressor fast enough to provide combustion and cooling air for the engine start and also turn the accessory section at a speed sufficient for the engine-driven fuel pump to deliver fuel to the combustor. Neither the compressor nor the starter by themselves have sufficient power to accelerate the engine from rest to idle speed, but when used in combination, the process takes place smoothly in about 30 seconds. The engine-start sequence is normally initiated by a switch in the cockpit but is often times terminated by a speed sensor that terminates the start sequence at a speed slightly above that at which the engine is capable of self-accelerating.

If an engine is not assisted to the speed at which it is capable of self-accelerating to idle speed, a *hung start* might occur. The engine will stabilize at a speed below or near starter cutoff speed, often with higher than normal temperatures. The engine must then be shut down and the cause of the hung start investigated. Any attempt to accelerate the engine at this point by adding fuel will likely result in a hot start as well as a hung start. There will be sufficient airflow for combustion but insufficient airflow for cooling purposes.

A typical starting sequence is initiated in the cockpit by beginning engine rotation. At speeds of 5 to 10 percent N_2 RPM, fuel will be introduced and ignition activated. Light-off should occur in 20 seconds or less (dependent upon engine type), and the engine should continue to accelerate and stabilize at idle RPM. Temperatures during start are closely monitored since any lag in acceleration or restriction to airflow through the engine can cause destructively high starting temperatures.

Airstart envelope

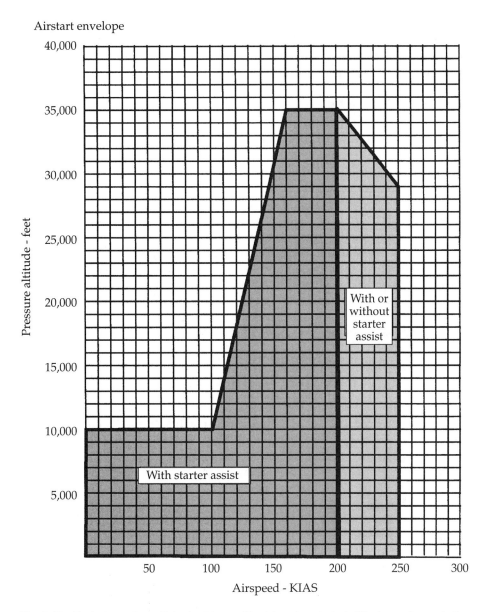

Fig. 4-18. Airstart envelope. This chart is used to determine at what altitudes and speeds the engine can be started in the air using starting assist or windmilling.

Starter-generators

The combination starter-generator is widely used on corporate-sized jets due to the weight savings involved in using one engine accessory for two purposes. This engine accessory receives electrical power from the aircraft batteries or external power source to rotate the engine for starting. When the start sequence is terminated, the starter-generator, which has a drive spline permanently engaged to the engine, reverts to a generator. This is typically accomplished through electrical switching. When the engine is rotating the starter-generator at sufficient RPM to produce full power, the generator can be connected to the aircraft electrical buses.

Pneumatic starters

The pneumatic, or air turbine, starter is a type of low-pressure air motor. This type of starter is used on almost all large commercial aircraft and some larger corporate aircraft. Its light weight, simplicity, and economical operation are an advantage over other starting methods.[15] The starter turbine transmits power to the starter output shaft through a reduction gear and clutch mechanism. The air required to spin the turbine might be obtained from an onboard auxiliary power unit (APU)[16], an external ground supply, or from a cross-feed pneumatic valve from a running engine. Air supply to the starter is typically regulated to begin at the initiation of the start cycle and be shut off at a predetermined starter speed.

The turbine of a typical pneumatic starter rotates up to 60,000 to 80,000 RPM and is geared down 20 to 30 times to achieve the high torque necessary to rotate the heavy, multistage axial compressors found on large commercial and military-transport aircraft. The clutch mechanism automatically disengages the starter from the engine as the engines reaches idling RPM.

ENGINE LUBRICATION

Oil consumption in gas-turbine engines is low compared to that of piston engines. A typical small-sized corporate jet with an oil capacity of three to five quarts will use one quart of oil every 100 to 300 hours.[17] Even the largest commercial airliners with oil tank capacities of 20 to 30 quarts use no more than 0.2 to 0.5 quart per hour. The demands made upon a lubrication system for an engine rotating at up to 40,000 RPM at temperatures of turbine combustion, however, are greater than those of piston engines.

The lubrication system is required to provide lubrication and cooling for all bearings, gears, and splines. It must also be capable of removing foreign matter that would cause rapid deterioration or failure of the engine if not removed. The oil must also protect those components of the engine that are not constructed on corrosion-proof materials, and it must do all these jobs without significant deterioration.

Because of the requirement to have a high enough viscosity for good load carrying capability, but also be of sufficiently low viscosity to provide good flow ability, most turbine oils are of the synthetic rather than petroleum-based variety. Synthetic oils used in turbine engines must be capable of operating over a wide temperature range of approximately –60°F to +400°F, and must not require preheat above –40°F. Additionally, the oil must retain its viscosity when heated to its operating temperatures and have excellent qualities of cohesion and adhesion allowing the oil molecules to stick together under compression loads and to stick to surfaces under rotational loads. Turbine oils must also have antifoaming properties and be resistant to evaporation at high altitudes.

Types of lubricating systems

Wet-sump

The wet-sump lubrication system is the oldest design and is rarely seen in modern flight engines. This type system is still found, however, in auxiliary power units and ground power units. The components of a wet-sump system are similar to a dry-sump system with the exception of the location of the oil supply. A wet-sump system does not have an oil tank and carries the oil supply in the engine accessory gearbox located at the lowest point in the engine.

Most wet-sump systems are variable-pressure type systems and do not incorporate a pressure-relief valve. The oil pump's output pressure is directly dependent upon engine revolutions per minute. A vent line in the gearbox prevents overpressurization and is vented to the atmosphere.

Dry-sump systems

Most gas-turbine engines utilize a dry-sump oil system that has pressure, scavenge, and breather-vent subsystems. An oil tank carries the main oil supply, and a smaller supply is carried in the accessory gearbox that also houses the oil pressure and scavenge pumps (Fig. 4-19).

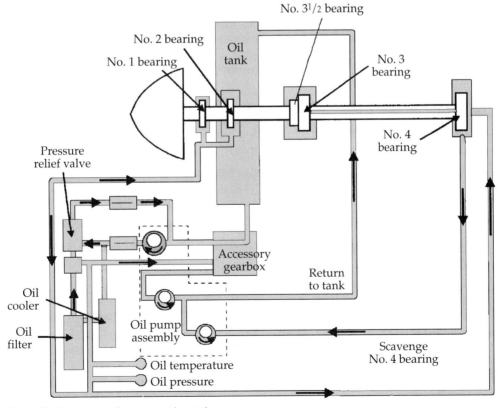

Fig. 4-19. Dry-sump oil system schematic.

System components

Oil tank The oil tank might either be a separate tank constructed of sheet metal and mounted on the engine or integral to the engine, as in a cavity between engine cases. Most oil tanks are designed to allow a pressure buildup of approximately 3-6 PSI above the ambient pressure to assure a positive flow of oil and inhibit foaming in the tank that could lead to oil-pump cavitation. Oil quantity is checked the same way in most gas-turbine engines as it is in piston engines, via the dipstick. Large commercial engines might also have oil-quantity gauges located in the cockpit.

Oil pumps The function of the oil pumps is to supply oil under pressure to the engine parts requiring lubrication. Many pumps are constructed with not only the pressure lubrication element but one or more scavenge pump elements in the same housing. Turbine-engine oil pumps might be of the vane, Gerotor, or gear-pump type, but all three types are positive, or constant-displacement pumps supplying a fixed quantity of oil with each revolution. These pumps are self lubricating and require positive pressure in the oil system to prevent cavitation and foaming.

Vane-type pumps are considered to be more tolerant of debris in scavenged oil and are lighter than the other two types of pumps, but might not have the mechanical strength of the Gerotor and gear-type pumps. Of the three types of pump, the gear-type is the most common.

Oil pumps are vital to the operation of the engine and are therefore not usually fitted with shear-type shafts found on other accessory components. The oil pumps must continue to supply oil pressure to the engine as long as the engine is rotating, regardless of the damage sustained by the pump. Loss of oil pressure necessitates rapid shutdown of the engine.

Oil filters Oil filters in gas-turbine engines must be capable of removing fine particulate matter from the oil to prevent damage to the engine and still allow free passage of oil over a wide range of operating temperatures. Contaminants found in oil system filters come from several sources:

- Dirt and other foreign matter introduced into the oil during servicing
- Metallic particles from corrosion and engine wear in the wetted areas of the engine
- Airborne contaminants entering through the main bearing seals
- Products of decomposition of the oil itself, usually seen as small black specks of carbon

Contaminants seen in filter bowls of turbine engines are always of concern. The fine tolerances and speed of rotation of turbine engines increases the potential for damage from any particulate matter. Determination as to whether particulate matter found renders the engine unairworthy is usually left to experienced maintenance personnel. Spectrometric oil analysis can determine the exact volume and proportion of metals in the contaminant and aid in the decision making.

Chip detectors Many engine-oil scavenge systems contain magnetic chip detectors that attract and hold ferrous metal particles that would otherwise circulate back to the oil tank and pressure system and potentially cause additional engine wear or damage. The chip detectors, which might be equipped with a cockpit warning device, are a point of frequent inspection. The presence of small fuzzy particles or gray metallic paste is considered an indication of normal wear. Metallic chips or flakes, however, are considered to be signs of serious internal wear or malfunction and might be early signs of main bearing failure. When the chip detector is equipped with a cockpit warning light, the illumination of the light will alert the flight crew to take the appropriate action, which might include inflight shutdown.

Oil coolers Various types of oil coolers are found in turbine engines, but the most common by far is the liquid-to-liquid oil cooler using the turbine fuel as the agent for cooling the oil. Most coolers have a chamber through which a series of pipes that are about the size of soda straws carry fuel. The oil is swirled around these straws and transfers its heat to the fuel. This arrangement, depending upon its placement in the engine fuel system, can also serve as a fuel heater.

Oil servicing

Mixing different brands of oil in turbine engines is usually either prohibited or greatly limited by engine manufacturers. Different brands of synthetic turbine oils have differing chemical compositions, and mixing might cause deterioration or foaming in the oil system. Most jet engines are placarded with the oil type being used. If oils are inadvertently mixed, the oil system must typically be drained and flushed and reserviced with the proper type oil.

The time of oil check and servicing is also important. Turbine-engine oil levels are checked within a short time of shutdown to prevent erroneous readings and overservicing of the oil supply. Oil added at each servicing is recorded to chart the normal oil consumption of the engine. Engine manufacturers specify both normal and maximum-allowable oil consumptions for each engine model. Steady oil consumption within allowable limits provides a valuable trend analysis to indicate that wear at main-bearing oil-seal locations is normal.

BLEED-AIR SYSTEMS

Bleed air is extracted from gas-turbine engines for various engine and airframe tasks. The engine internal-air system is defined as those airflows that do not directly contribute to engine thrust. The system has several important functions to perform for the safe and efficient operation of the engine including internal engine and accessory unit cooling, bearing chamber sealing, control of bearing axial loads, and engine anti-icing. Additionally, bleed air is supplied for aircraft services such as cabin pressurization, air conditioning, and surface anti-icing systems. Up to 20 percent of the total engine core mass airflow may be extracted for these various functions.

Bleed-air extraction

As air moves through the compressor stages, increasing amounts of work are done upon it to raise its pressure and temperature. In order to reduce engine performance losses, air for internal and aircraft service needs is taken as early in the compression process as practical commensurate with the requirements of each specific function. The source of bleed air is typically identified with the compressor stage at which extraction takes place, for instance, P_3 bleed air will be extracted from the third compressor stage. Occasionally, during some phases of the flight cycle, it might be necessary to switch the bleed-air source for a specific function, typically cabin service needs, to a later stage of compression to maintain adequate temperature and pressure.

Bleed-air extraction for constant-use items such as turbine and accessory cooling, bearing sealing, control of bearing axial loads, and cabin pressurization is normally taken into account in the fuel scheduling of the engine. This air is deducted from the mass airflow through the core of the engine before computing the fuel schedule to the combustor. Depending upon the sophistication of the fuel-control unit, intermittent-use items such as engine and surface anti-ice bleed-air extraction might be automatically compensated for by the fuel-control unit or manually adjusted by the pilot. The aircraft flight manual and checklists will contain information to assist the pilot in this operation.

Table 4-2 shows a partial climb-power-setting table for the JT15D engine installed on Cessna Citation aircraft showing both anti-ice off and anti-ice on power settings. Notice that the power settings for anti-ice on are in most cases lower than those without this bleed extraction. There is no input to the fuel-control unit on this engine installation to compensate for engine anti-ice bleed extraction; therefore; the pilot must reduce the power setting to compensate for the loss in cooling air to the engine.

Table 4-2. Climb power settings - JT15D-4 Turbofan engine
Normal Climb Thrust Setting - N1 % RPM

Pressure Altitude Feet	Ram Air Temperature—°C								
	15	10	5	0	−5	−10	−15	−20	−25
Sea	97.2[1]	96.6	95.8	95.2	94.4	93.8	93.0	92.4	91.6
level	—	95.2[2]	95.8	95.2	94.4	93.8	93.0	92.4	91.6
5,000	97.8	98.9	100.0	101.2	102.4	102.2	101.4	100.6	99.7
Feet	—	95.2	96.0	96.9	97.7	98.4	99.0	99.6	99.7
10,000	97.8	98.9	100.0	101.2	102.4	103.4	104.0	104.0	103.6
Feet	—	95.2	96.0	96.9	97.7	98.4	99.0	99.6	100.0
15,000	97.8	98.9	100.0	101.2	102.4	103.4	104.0	104.0	104.0
Feet	—	95.2	96.0	96.9	97.7	98.4	99.0	99.6	100.0
Above	97.8	98.9	100.0	101.2	102.4	103.4	104.0	104.0	104.0
20,000	—	95.2	96.0	96.9	97.7	98.4	99.0	99.6	100.0

1. Upper Value for Anti-Ice OFF
2. Lower Value for Anti-Ice ON

Engine internal-air systems
Turbine cooling

Turbine cooling is necessitated by the operating environment of the turbine components. High thermal efficiency of a gas-turbine engine is dependent upon high turbine entry temperatures, which are limited by the turbine blade and guide vane materials. Continuous cooling of these components allows their environmental operating temperature to exceed the material's melting point without affecting the blade and vane integrity. Heat conduction from the turbine blades to the turbine disc requires the discs to be cooled to prevent thermal fatigue and uncontrolled expansion and contraction rates (Fig. 4-20).

Accessory cooling

Some engine accessories, notably the electrical generators, produce a considerable amount of heat during their normal operation and often require their own cooling circuit. When air is used for cooling, the source might be compressor air or atmospheric air ducted from intakes in the engine cowling. When atmospheric air is used during flight to achieve this cooling, it is usually necessary to provide some means to induce adequate airflow during static ground operations. This is achieved by routing compressor air through nozzles, or ejector pumps[18], in the cooling-air outlet ducting and so inducing an airflow of cooling air through the nacelle intakes.

Sealing

Seals are used to prevent oil leakage from engine bearing chambers to control cooling airflows and to prevent the combustion chamber discharge gases from entering the turbine disc cavities. Various sealing methods are used in gas-turbine engines depending upon the surrounding temperature and pressure, wearability, weight, space available, and heat generation.

Labyrinth seals and carbon seals are among the most common sealing methods used in gas-turbine engines. Labyrinth seals are widely used to retain oil in bearing chambers. Carbon seals are used on rotating shafts and rely on the oil system to dissipate the large amount of heat generated by this sealing method. Figure 4-21 illustrates labyrinth and carbon seals.

Control of axial-bearing loads

Turbine engine shafts experience varying axial loads that act in a forward direction on the compressor and in a rearward direction on the turbine. Because of these loads, the shaft connecting these components is always under tension. The difference between the fore and aft loads is carried and balanced by a location bearing that is kept adequately loaded throughout the engine thrust range by the force of internal air pressure on a fixed-diameter balance seal.

Engine anti-ice

Icing of the leading edges of the engine intake duct and the first stage stators can take place during flight through clouds containing super-cooled water droplets or

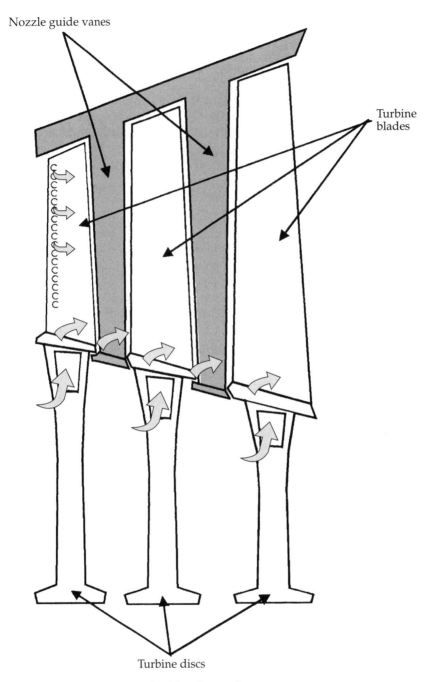

Nozzle guide vanes

Turbine blades

Turbine discs

Fig. 4-20. Turbine-blade and turbine-disc cooling.

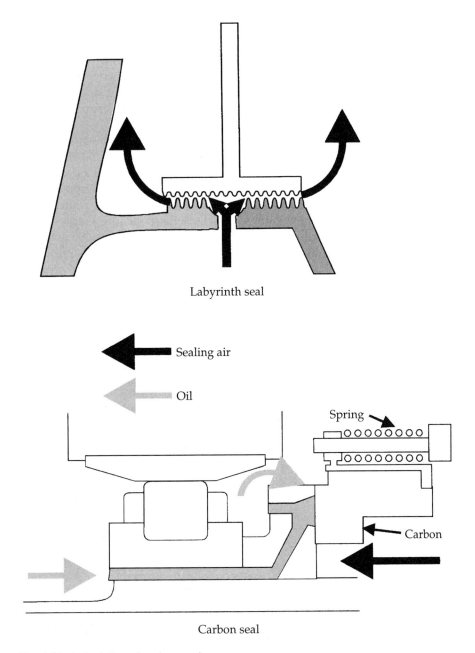

Labyrinth seal

Sealing air

Oil

Spring

Carbon

Carbon seal

Fig. 4-21. Labyrinth and carbon seals.

during ground operations in freezing fog. Additionally, ice can form in the engine inlet at temperatures up to 45°F due to the temperature drop of the air mass flowing into the inlet. This is similar to the venturi effect causing carburetor ice in piston engines. Protection against icing must be provided since icing of these regions can restrict or disturb the airflow through the engine causing loss in performance and possible compressor stalls or other engine malfunctions. Additionally, damage might result from ice breaking away and being ingested into the engine.

An ice protection system must be effective in preventing ice formation within the operational limits of the aircraft. Additionally, the system must be reliable, easy to maintain, present no excessive weight penalty, and cause no serious loss of engine performance when in operation.[19]

Because a gas-turbine engine always has excess air available, the most common source of heat for engine anti-icing systems is compressor bleed air. This air is extracted from the high-pressure compressor or diffuser case at a point that will cause the least loss of engine performance but provide adequate temperature and pressure to accomplish its purpose (Fig. 4-22). Bleed air is then routed through regulator valves and then to the engine inlet cowling and the first set of stator vanes and/or inlet guide vanes. The nose cone of the fan is typically continuously anti-iced by an unregulated bleed-air supply. Exhaust air from the anti-icing system might be routed back into the primary airstream or exhausted overboard. Cockpit controls for engine anti-ice systems vary from totally automatic to electri-

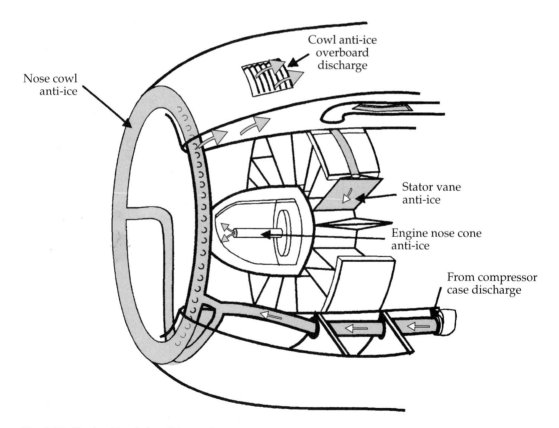

Fig. 4-22. Engine bleed-air anti-ice system.

cal switches and manual valves. Checklists and aircraft flight manuals will outline the normally simple procedures necessary for engine anti-icing, but the importance of the system should never be diminished because of its simplicity.

FIRE PREVENTION, DETECTION, AND PROTECTION

Fire prevention

All gas-turbine engines incorporate features that minimize the possibility for uncontained combustion and ignition of flammable liquids and engine parts and accessories. In most instances, a dual failure is necessary before a fire can occur. Most flammable liquids are isolated from the "hot end" of the engine. External fuel and oil components and their associated lines are usually located around the compressor casings in a cool zone of the engine and are separated from the combustion, turbine, and jet pipe area by a bulkhead. Additionally, these areas can be ventilated by atmospheric air to prevent accumulation of flammable vapors.

All pipes and lines that carry flammable liquids such as fuel, oil, and hydraulics are constructed of fire-resistant/fireproof materials and, in many cases, are constructed of double-walled materials. Additionally, all electrical components are explosionproof and are bonded to the aircraft structure to prevent buildup and discharge of static charges and/or arcing.

Fire containment

Any engine fire that might occur must be contained within the powerplant and not be allowed to spread to other parts of the aircraft. Although the engine nacelles on most gas-turbine engines are constructed of aluminum alloys that would be incapable of containing a fire in the static condition, the situation while in flight is much different. During flight, airflow around the nacelles provides sufficient cooling to render them fireproof. Fireproof bulkheads and cowlings that do not have the benefit of this cooling airflow are constructed of steel or titanium.

Fire detection

Fire-detection systems must reliably detect the presence of engine fires without giving false alarms due to chafing, electrical shorts, or other system malfunctions. Several different methods of fire detection are found on gas-turbine engines, most of which do not actually detect the presence of fire, but only detect excess heat in the engine area. The most common detection systems operate based upon changes in electrical resistance with increasing heat or the uneven expansion of a bimetallic disc when heated or a pneumatic detection system based upon the expansion of gases with heat. A common temperature limit for fire-detection systems is 500°F since the combustion section of a typical engine normally radiates heat of 300°F or slightly more. Whichever means is used to detect the fire, the system will activate a cockpit warning system and alert the crew to necessary action.

Fire extinguishing

Although the actions necessitated by an engine-fire warning will be dictated by the aircraft flight manual, several items have been found to be common across platforms:

1. The throttle is normally retarded on the affected engine to preclude the possibility that the engine fire warning has been triggered by excess heat in the nacelle area caused by a compressor bleed-air leak. If this is the case, reduction of the throttle might well reduce the temperature of the bleed air below the trigger threshold.

2. If throttle reduction has not eliminated the fire warning, it is necessary to stop the engine prior to fire extinguishing to reduce the discharge of flammable liquids and air into the fire area and to prevent the rapid dissipation of the extinguishing agent. This is normally accomplished by a firewall shutoff switch or handle located in the cockpit.

3. Once the engine has been stopped by actuation of the firewall shutoff, the extinguishing agent can be introduced into the engine.

4. Once the fire has been extinguished, no attempt must be made to restart the engine. To do so would risk reestablishing the fluid leak and the ignition source that were the original causes of the fire.

Several extinguishing agents have been used in gas-turbine engines, the oldest of which is carbon dioxide. This gas is noncorrosive to engine parts, but can cause shock to hot-running engine parts if used in large quantities. Dibromodifloromethane (Cbr_2F_2) and trifluorobromomethane (CF_3Br) are both much more expensive than other extinguishing agents, but are noncorrosive to engine parts and do not require that the engine be washed down after discharge of the agent. Both agents are very affective against engine fires.

The extinguishing agents are contained in pressurized containers located either in the engine compartment away from the fire-risk zone or the airframe near the engines. Typical container pressure is 600 PSI and is somewhat variable with ambient temperature. Cockpit controls and switches control the discharge of the extinguishing agent into the engine. Fire-extinguishing systems generally enable two separate discharges to be made into the engine area. Figure 4-23 illustrates a typical fire-detection/protection system.

Engine overheat detection

Although turbine overheat does not constitute a serious fire risk, detection of overheat conditions is necessary to enable the pilot to stop the engine before mechanical damage results. Warning circuits similar to the fire detection circuits can be used for this function as can thermocouples positioned in the cooling airflow. Pilot response to these warnings will vary from power reduction to engine shutdown and will be dictated by the aircraft flight manual.

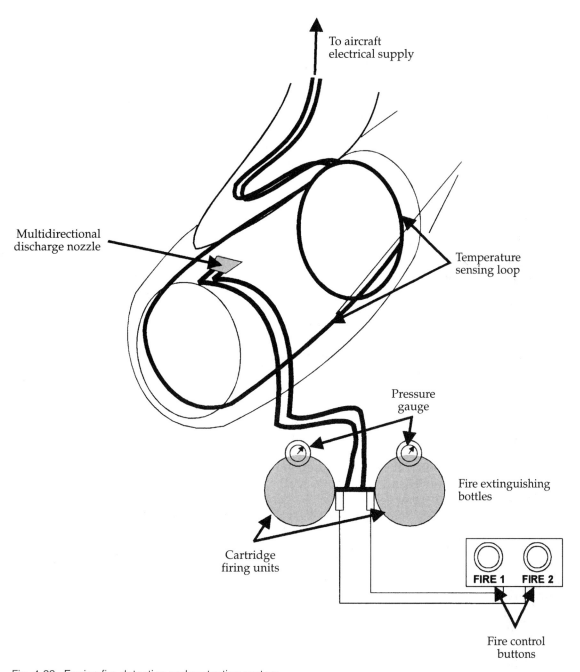

To aircraft
electrical supply

Multidirectional
discharge nozzle

Temperature
sensing loop

Pressure
gauge

Fire extinguishing
bottles

Cartridge
firing units

FIRE 1   FIRE 2

Fire control
buttons

Fig. 4-23. Engine fire detection and protection system.

THRUST REVERSAL

One of the most common observations made by new jet pilots is the reluctance of a typical jet to stop on the runway. The elimination of windmilling propellers eliminates more drag than the pilot is usually aware exists. Jets do not have these large rotating airbrakes to assist deceleration on the runway, and the immediate application of heavy wheel braking certainly shortens the life of the brake components. Wet, icy, or snowy runways might further reduce braking efficiency by a loss of adhesion between the aircraft tire and the runway. Thrust reversal is the answer to these dilemmas and uses a nonconsumable resource to stop the aircraft. Additionally, the elimination of residual thrust from the engine aids deceleration. Most jet aircraft can be felt to accelerate slightly when thrust reversers are stowed due to this residual thrust.

Thrust reversal can be accomplished with either bypass air, primary air, or a combination of both. Ideally, the gases should be directed in a completely forward direction; however, it is not possible to achieve this, mainly for aerodynamic reasons. The gases are therefore discharged at an angle of approximately 45° forward, and reverse thrust is consequently somewhat less than the forward thrust obtained for the same throttle displacement.

The two most common arrangements for reversal of thrust are the *clamshell-door* system and the *target* or *bucket* system. The clamshell arrangement is illustrated in Fig. 4-24, and the target type system is shown in Fig. 4-25. The clamshell doors are normally operated pneumatically. Normal engine operation is not affected by the system because the ducts through which the gases are deflected remain closed by the doors until reverse thrust is selected by the pilot. Normal air loads in the engine in the forward-thrust position ensures effective sealing of the door edges to preclude gas leakage and degradation of performance. The target-type or bucket reversers are normally hydraulically actuated and provide a barrier behind the engine to deflect the exhaust gases. The thrust-reverser doors provide the final stage of the convergent-divergent exhaust nozzle for the engine.

Methods of reverse thrust selection and the safety features built into each system are fundamentally the same no matter what type system is installed. Reverse thrust is selected by the pilot by a reverse-thrust lever in the cockpit. The reverse-thrust lever usually cannot be moved to the deploy position unless the aircraft is on the ground and the engines are at idle or near-idle power settings, and the engine cannot be accelerated to a higher power setting until the reverser doors have moved to the fully deployed position.

Most reverser systems require that the engine be reduced to idle thrust at a predetermined minimum ground speed to preclude ingestion of foreign objects and hot exhaust gases into the engine inlet. Additionally, FAR 25 requires that if a method of reversing thrust is installed on an aircraft and reverse thrust is not authorized in the air, then that thrust reversal system must have a means of immediately reducing the engine power to idle in the event of an inadvertent inflight deployment of the reverser, or the aircraft must have been found to have acceptable flying characteristics with the reverser deployed and the engine at a higher power setting. Operation of the reverser system is annunciated to the pilots by a set of cockpit indicator lights.

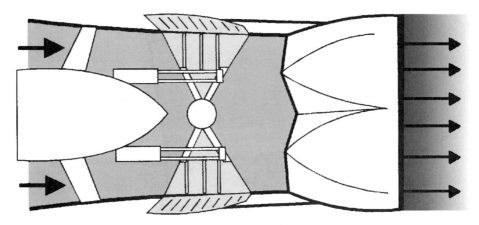

Clamshell doors in forward
thrust position

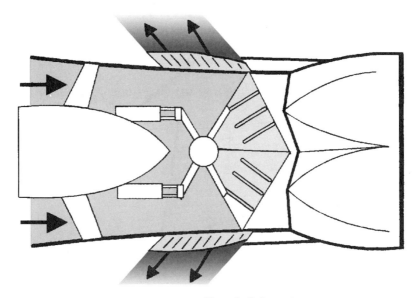

Clamshell doors in
reverse thrust position

Fig. 4-24. Clamshell thrust reversers.

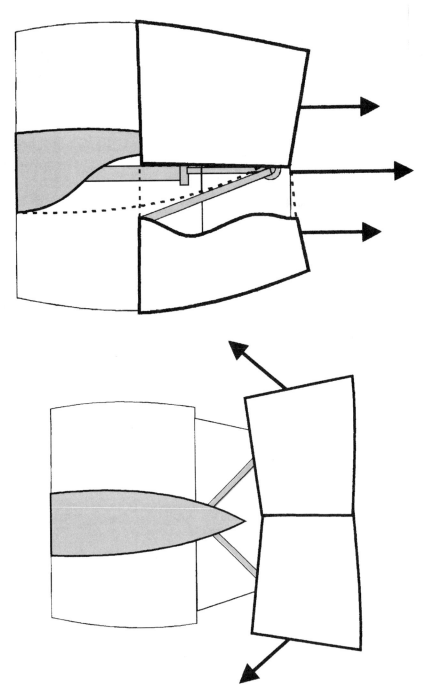

Fig. 4-25. Target, or bucket, thrust reversers.

GAS-TURBINE ENGINE INSTRUMENTATION

Many of the engine indicating instruments found in jet cockpits are the same or very similar to those found in piston aircraft. Familiar instruments include gauges for oil pressure and temperature (and in some cases oil quantity), fuel quantity and pressure and flow, and exhaust gas temperatures. Only the temperature and quantity ranges might be different. Other gauges found in the jet cockpit, however, are totally foreign to the piston pilot. Such things as N_1 and N_2 tachometers, EPR or engine pressure ratio gauges, ITT gauges, and engine vibration gauges will not look at all familiar.

N_1 and N_2 tachometers

Turbine engines have a tachometer for compressor or "spool." N_1 will usually refer to the fan stage of the engine (if any), and the highest number will be assigned to the highest-pressure compressor. The major difference between the tachometers on piston aircraft and those on gas-turbine engines is that the gas-turbine engine tachometer is calibrated in percent RPM. Because no two compressors, N_1 and N_2, operate at exactly the same speed, percent revolutions per minute is used to simplify the cockpit indications.

Percent tachometers are used for the following engine monitoring operations:

- Monitor engine RPM and acceleration during start sequence
- Monitor engine RPM in case of overspeed
- Indicate both compressor speed and thrust on some turbofan engines (Refers to N_1 speed.)
- Monitor operating schedules of engine support systems such as anti-ice, bleed valves, variable vanes, and the like
- Monitor engine speed and deceleration during engine shutdown

Percent RPM tachometers are driven by various means, the two most common being a mechanical tachometer generator and a magnetic pickup. The mechanical tachometer generator is an independent electrical system consisting of an ac generator driven by a tower shaft from the appropriate turbine shaft and a synchronous motor-driven indicator. Engine revolutions turn the shaft of the tachometer generator and these are converted by the indicator into percent RPM. The magnetic pickup is mounted in the fan case and counts blade-tip passage and translates this into percent RPM.

No matter what method is used to measure percent RPM, these gauges are used to monitor engine limitations. Overspeeds of any turbines in the engine place unacceptable stress on these parts and must be avoided. Any RPM overspeed exceeding limitations set by the engine manufacturer must be recorded in engine records, and the appropriate maintenance action must be performed. This maintenance can range from simple inspection to engine removal and overhaul depending upon the magnitude and duration of the overspeed.

Percent N_1 RPM is used as the primary power setting gauge for maximum allowable power in many turbofan engines. This might be used on its own or with an engine-pressure-ratio (EPR) gauge as a backup. Although power response in a piston engine is directly proportional to engine RPM, the power response of a turbofan is not so straightforward. Thirty percent of the power developed by the turbofan engine is developed in the top 10 percent of allowable RPM. Keep this in mind the next time you hear a jet crew making a large power reduction soon after takeoff as mandated by some airport noise-abatement takeoff regulations.

Engine-pressure-ratio (EPR) gauges

Engine-pressure ratio (EPR) has been the most widely used thrust indicating system for turbojet cockpits. It is still used on many turbofan engines as a primary thrust setting instrument and as a condition-monitoring instrument on some turbofan engines that utilize N_1 speed as the primary performance instrument.

EPR gauges simply indicate the ratio of the turbine discharge pressure to the compressor inlet pressure. Recall that a gas-turbine engine is a device that increases potential energy in the form of high-pressure gas and then converts this potential energy into kinetic energy in the form of a high-velocity jet of gases. These gases exiting the tail pipe create a reacting force known as thrust. The total pressure in the tail pipe represents the force it would take to completely stop the flow of gases. With the relatively high velocities of the gases and the relatively large discharge area of the tail pipe, it is easy to see how gas under rather low pressure (up to 2.5 times ambient) can have a rather large force (thrust) value.

Each engine manufacturer uses a slightly different method of labeling engine stations, but the letter P with a subscript number generally indicates the engine station for the pressure tap-off points. Pratt & Whitney, for example, uses stations 2 and 7, P_{t2} and P_{t7}, to identify the engine-pressure-ratio tap-off points in its dual-spool engines; therefore, if the turbine discharge pressure is 28.96 pounds per square inch—absolute and compressor inlet pressure is 14.7 pounds per square inch—absolute, the EPR would be 1.97, calculated as follows:

$$EPR = P_{t7}/P_{t2}$$
$$P_{t2} = 14.7 \text{ psia}$$
$$P_{t7} = 28.96 \text{ psia}$$
$$EPR = 28.96/14.7 = 1.97$$

ITT/TIT/EGT/TOT gauges

The temperature of the exhaust gases in a gas-turbine engine is always closely monitored, especially during the start cycle when overheat damage is most prevalent. TIT (turbine inlet temperature) is considered the most critical of all engine operating parameters because an out-of-limits condition can render the engine unairworthy in a matter of seconds. ITT (interstage turbine temperature), which refers to a temperature reading taken at some intermediate stage between turbine wheels, TOT (turbine outlet temperature), and EGT (exhaust gas temperature) are all various means of measuring and monitoring basically the same parameter.

Regardless of the actual monitoring process, the TIT just in front of the first turbine inlet nozzle is the crucial consideration. It is not always practical to position a temperature probe or probes in this position, however, because the high operating temperatures drastically shorten the service life of temperature-sensing probes. Manufacturers therefore have provided a comparison value between the actual TIT and the measurements taken elsewhere in the engine. The temperature of the exhaust gases, no matter how they are measured or displayed, indicates the integrity of the turbine components more than any other engine-instrument system.

Every engine has a strict set of internal temperature limits. Normally, the limits for starting are more stringent than operating limitations. The sudden temperature rise during engine start is more destructive than higher operating temperatures in flight. Table 4-3 sets forth a typical set of engine-operating temperature limits and the actions required when limitations are exceeded.

Table 4-3. Engine temperature limitations

Temperature	Duration	Action
Starting Overtemperature Limitations		
Up to 525°C	—	None (normal)
526 to 631°C	5 sec or less	Determine cause and correct
526 to 594°C	Any length of time	Determine cause and correct
595 to 630°C	Over 5 sec	Visual inspection of hot section
631 to 700°C	5 sec or less	Visual inspection of hot section
631 to 700°C	Over 5 sec	Teardown inspection of hot section
701 to 800°C	Any length of time	Teardown inspection of hot section
801°C	Any length of time	Complete overhaul inspection of hot section
Operating Overtemperature Limitations		
Up to 677°C	—	None (normal)
678 to 720°C	5 sec or less	Determine cause and correct
678 to 690°C	Any length of time	Determine cause and correct
721 to 730°C	Over 5 sec	Visual inspection of hot section
691 to 720°C	Over 5 sec but less than 2 mins	Visual inspection of hot section
690 to 720°C	Over 2 mins	Teardown inspection of hot section
721 to 730°C	Over 5 sec	Teardown inspection of hot section
731 to 775°C	Any length of time	Teardown inspection of hot section
776°C	Any length of time	Complete overhaul inspection of hot section

Engine vibration gauges

Turbine engines operate at extremely low levels of vibration due to the fine tolerances and relatively small number of moving parts and the consistent movement (rotation) of all moving parts in the same direction.[20] Consequently, a small increase in vibration might be unnoticeable by the crew but its detection might be crucial in forestalling major engine damage or catastrophic failure. Engine vibration gauges normally show the vibration amplitude in mils (thousandths) of an inch measured at four locations on the engine: at the fan and low-pressure turbine and at the compressor and high-pressure turbine. Four mils (4/1000 inch) are normally the maximum allowable vibration at any single point in the engine.

Endnotes

1. To be absolutely correct for all locations in the universe, weight is equal to the product of the mass of an object in grams, or kilograms, and the acceleration due to gravity in meters per second per second and is expressed in Newtons (N). 1 N = 1 kg × 1 m/s². 1 lb. = 4.448 N. Because weight depends on gravity and gravity varies with the distance from the center of the Earth, weight will vary with altitude. I hope you will agree with me that for this discussion we can ignore such exactitude.

2. Or, stated in a manner more likely to please physicists, temperature is a direct measure of the average molecular kinetic energy of an ideal gas.

3. For the physicists, the pressure of a gas is proportional to the number of molecules per unit volume and to the average translational kinetic energy of the molecules.

4. Strictly speaking, the work done, W, by a constant force is defined as the product of the component of the force along the direction of displacement and the magnitude of the displacement.

5. Scalar quantities, while they may be identified with units, are quantities that might be specified with magnitude alone, without regard to direction. Examples are temperature, mass, volume, speed, length, and time.

6. When a bullet is fired, work is done on the bullet to put it into motion. When the bullet strikes an object, it does "work" on that object. Work is also done on the bullet to bring it to rest.

7. Air is, of course, compressible, but as regards to flow in a duct, it is not. In other words, the air will change its static or dynamic pressure in regards to changes in the duct but will not "build up" or compress.

8. Interestingly, damage that results from internal failure of engine parts is termed DOD or domestic object damage.

9. The total pressure will remain the same if the mass airflow does not change. Only static and dynamic, or ram, pressures will change their values.

10. These similarities in basic systems is why you might have heard that your first type rating is the hardest; the rest are much simpler.

11. This will sometimes be expressed as 60:1 air-to-fuel ratio. In that case, the ratio is based on the total air being routed to the jet core, not just the primary air actually used in the combustion process.

12. Fuel control units are based on weight; therefore, turbine-airplane fuel gauges are calibrated in pounds. The carburetor and fuel-injection systems found on piston aircraft meter fuel based on volume, which is why their gauges are calibrated in gallons.

13. Some engines such as the Pratt & Whitney JT15D series found on the Cessna Citation 500 series aircraft and the Beechjet have oil coolers utilizing fuel to cool the lubricant; however, due to the position of the oil cooler downstream from the fuel control unit, it cannot be considered a fuel heater.

14. Prior to engine light-off, it is possible to hear the igniters "snapping" as they spark in the burner can. I have had pilots swear to me they can hear the difference between one igniter and two, but they surely have better perception than I because try as I might, I cannot tell the difference.

15. Electric starters, explosive-cartridge starters, and hydraulic starting systems have all been used at one time or another, but most smaller corporate jets use starter-generators with the larger corporate and commercial craft using pneumatic starters.

16. Auxiliary power units, or APUs, are nothing more than a small gas-turbine engine located in the fuselage of the aircraft, usually the tail section. One APU used on some models of 747 aircraft is the Pratt & Whitney JT15D turbofan—the same engine that powers the Citation 500 aircraft and the Beechjets.

17. This must be mostly the amount of oil that is dripped from the dipstick when checking the oil since turbine engines do not consume oil in the same way that piston engines do. Oil is not burned in turbine engines in the process of lubricating cylinder piston rings, nor is it blown from exhaust valves. Oil systems in turbine aircraft are not in contact with the combustion process.

18. These ejector pumps are similar to and operate under the same principles as those described in the fuel system. These pumps, all operating as venturis according to Bernoulli's theorem, are found in many systems on jet aircraft.

19. As we have seen from the power-setting table presented above for the Citation II aircraft, there is some performance loss with the activation of engine anti-ice, but it must be remembered that these systems are used at ambient temperatures where relatively small performance losses are not crucial.

20. Jet aircraft have so little inherent vibration that the altimeters installed on jets have small internal vibrator motors installed to keep the altimeter hands from sticking in place. This certainly never was a requirement on the Beech 18s that I flew!

5

Environmental systems

AS DISCUSSED IN CHAPTER 3, WE PILOTS ARE VERY DEPENDENT UPON our fragile cocoon and its contained environment. There must be a means for providing air at a sufficient pressure to ensure proper oxygenation of the blood[1], and that air must be provided at a comfortable temperature. Pressurization and air conditioning might seem to be separate subjects, but they are actually two different facets of the same task. The aircraft is pressurized through operation of the air-conditioning system.

The environmental systems on jet aircraft seem to be among the systems that give most new jet pilots the most trouble in their studies. I don't believe that this occurs because these systems and concepts are particularly hard to understand, but more that these systems are among those that are most different on jet and other turbine-powered aircraft. These are also systems that do not get noticed at all if they are working properly. And, as an LA-area pilot once told me: "I never could trust air I can't see."

PRESSURIZATION

Pressurization did not become practical until there was a means of providing sufficient quantities of air at sufficient pressure to the cabin of the aircraft. Pressurization did not become necessary until passenger aircraft began flying higher in the atmosphere to escape the effects of low-level weather and to extend range. The turbosupercharger made both high flight and pressurization possible. The turbo-superchargers on the large radial engines of the Douglas DC-6 provided the denser air needed by the engine to produce power at high altitude and also provided the air necessary to pressurize the cabin.

Differential

All pressurized aircraft have a specified maximum pressurization differential, often referred to in pilot shorthand as "max diff." Many pilots use this term, but an amazing number of those I have questioned cannot adequately define maximum differential or pressure differential. Maximum differential is simply the maximum difference in pressure, in pounds per square inch, or PSI, that is allowed to exist between the interior of the cabin and the ambient outside pressure. It follows that pressure differential is whatever difference exists between inside and outside pressure, whether at the maximum or not.

Table 5-1 shows the pressure, in PSI, for the standard atmosphere at varying altitudes. If an aircraft has a maximum pressure differential of 9.4 PSI (as does the Learjet Model 36), then the pressure within the cabin at maximum differential will be 9.4 PSI higher than that outside the aircraft. If the aircraft altitude is known, the cabin altitude can be calculated with this information.

Table 5-1. International standard atmosphere

Altitude (feet)	British units (pressure)	
	PSIA	In. Hg.
0	14.70	29.92
1,000	14.17	28.86
2,000	13.66	27.82
3,000	13.17	26.82
4,000	12.69	25.84
5,000	12.23	24.90
6,000	11.78	23.98
7,000	11.34	23.09
8,000	10.92	22.23
9,000	10.50	21.39
10,000	10.11	20.58
15,000	8.29	16.89
20,000	6.754	13.75
25,000	5.45	11.10
30,000	4.36	8.89
35,000	3.46	7.04
40,000	2.72	5.54
45,000	2.14	4.34
50,000	1.68	3.43
60,000	1.04	2.12

Enter the table at the aircraft altitude, 45,000 in this case. The standard atmosphere ambient pressure at 45,000 is 2.139 PSI. Add the pressurization differential of 9.4 PSI to this figure, and you find the pressure inside the cabin to be 2.139 + 9.4 or 11.539 PSI. Referring again to Table 5-1, you will find that this pressure is found at an altitude between 6000 and 7000 feet, and this will be the cabin altitude. This calculation does not need to be made during flight because there is a handy little gauge on the panel of all pressurized aircraft that shows cabin altitude.

Pressure vessel

Before we can discuss pressurization, we need to know what we intend to pressurize. Not all parts of an aircraft are pressurized. The area that is pressurized is known as the pressure vessel. This extends from the forward pressure bulkhead, usually located just in front of the cockpit instrument panel, to the rear pressure bulkhead in the tailcone. Nosecone avionics compartments, nose baggage compartments, and tailcone access compartments are not normally pressurized.

Aircraft structures must be able to withstand repeated expansion and contraction in the highly variable temperature range found between sea level and the flight levels.[2] (A 747 "grows" several inches in diameter when pressurized at altitude.) Windows in a pressurized aircraft are doublepaned to provide an element of redundancy and to reduce fogging from the temperature and humidity changes encountered when climbing to and descending from altitude.

Doors are either of the plug type that cannot open while the aircraft is pressurized or are fitted with heavy pins that engage the door frame when latched to hold them closed against the pressure differentials that are developed. Doors are fitted with inflatable door seals that use regulated engine bleed air to seal the inevitable gaps between the door and its frame.

HOW PRESSURIZATION IS ACCOMPLISHED

There are two available methods of accomplishing aircraft pressurization; either assume a constant inflow of air and regulate the outflow, or assume a constant outflow and regulate the inflow. Because a steady supply of regulated bleed air can be extracted from the compressor case of a jet engine, the easiest method of pressurization is to assume a steady inflow of air and regulate the outflow.

Federal Aviation Regulation 25 sets forth requirements for controls and limitations for pressurized aircraft:

25.841 Pressurized cabins.

(a) Pressurized cabins and compartments to be occupied must be equipped to provide a cabin pressure altitude of not more than 8,000 feet at the maximum operating altitude of the airplane under normal operating conditions. If certification for operation over 25,000 feet is requested, the airplane must be able to maintain a cabin pressure altitude of not more than 15,000 feet in the event of any reasonably probable failure or malfunction in the pressurization system.

(b) Pressurized cabins must have at least the following valves, controls, and indicators for controlling cabin pressure:

(1) Two pressure relief valves to automatically limit the positive pressure differential to a predetermined value at the maximum rate of flow delivered by the pressure source. The combined capacity of the relief valves must be large enough so that the failure of any one valve would not cause an appre-

ciable rise in the pressure differential. The pressure differential is positive when the internal pressure is greater than the external.

(2) Two reverse pressure differential relief valves (or their equivalents) to automatically prevent a negative pressure differential that would damage the structure. One valve is enough, however, if it is of a design that reasonably precludes its malfunctioning.

(3) A means by which the pressure differential can be rapidly equalized.

(4) An automatic or manual regulator for controlling the intake or exhaust airflow, or both, for maintaining the required internal pressures and airflow rates.

(5) Instruments at the pilot or flight engineer station to show the pressure differential, the cabin pressure altitude, and the rate of change of the cabin pressure altitude.

(6) Warning indication at the pilot or flight engineer station to indicate when the safe or preset pressure differential and cabin pressure altitude limits are exceeded. Appropriate warning markings on the cabin pressure differential indicator meet the warning requirement for pressure differential limits and an aural or visual signal (in addition to cabin altitude indicating means) meets the warning requirement for cabin pressure altitude limits if it warns the flight crew when the cabin pressure altitude exceeds 10,000 feet.

(7) A warning placard at the pilot or flight engineer station if the structure is not designed for pressure differentials up to the maximum relief valve setting in combination with landing loads.

(8) The pressure sensors necessary to meet the requirements of paragraphs (b)(5) and (b)(6) of this section and 25.1447(c), must be located and the sensing system designed so that, in the event of loss of cabin pressure in any passenger or crew compartment (including upper and lower lobe galleys), the warning and automatic presentation devices, required by those provisions, will be actuated without any delay that would significantly increase the hazards resulting from decompression.

A constant flow of air under pressure enters the aircraft cabin through the underfloor and overhead vents. This air is provided by the air-conditioning system (to be subsequently discussed) but is essentially compressor bleed air that has been modified in temperature. No oxygen is added to this air. The air entering the engine inlets, and consequently the compressor, at altitude still contains the constant 21-percent oxygen; however, recall from chapter 3 that this air has insufficient pressure to oxygenate the blood. The job of the aircraft pressurization system is to raise the pressure of the air in the cabin to a level compatible with useful consciousness and to maintain that pressure throughout the authorized altitude range of the aircraft.

Air is perhaps one of the hardest substances to seal into or out of any container, and the existence of a difference in pressure between the inside and the outside of the container complicates this problem. Air cannot just be pumped into an aircraft cabin on takeoff and be sealed there for the duration of the flight. Not only would the air so sealed become stagnant and oxygen poor, but the pressure differential between the inside and outside of the aircraft could not be adjusted.

More pertinent, however, is the fact that the air will not remain in the cabin. All aircraft leak. The necessity of passing control cables and electrical wires through pressure bulkheads makes a hermetically sealed aircraft impossible. The average fleet leak rate of a popular corporate jet is 2000 to 2500 feet per minute. That is to say that if the source pressurization air is eliminated, the cabin will "rise" or equalize with the outside pressure, at a rate equal to a climb of 2000 to 2500 feet per minute. This leak rate, of course, is dependent upon variables such as the beginning pressure differential and the aircraft age and general condition, but the average rate is typical for this class aircraft.

The pressurization controller, by modulation of the cabin outflow valve(s), regulates the rate at which the pressurized air is allowed to escape the cabin and therefore regulates the cabin altitude. The cabin altitude, differential pressure, and cabin rate of climb/descent are shown on cockpit gauges (Fig. 5-1).

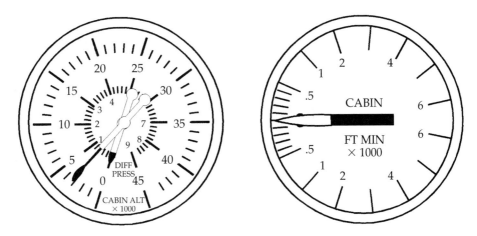

Fig. 5-1. The cabin altitude and differential pressure plus the cabin rate-of-change gauges allow the crew to monitor the pressurization system.

Outflow valves may be operated by pneumatic pressure (vacuum) or by electric motors. Outflow valves, which can be located at either or both ends of the pressure vessel, are modulated to maintain the desired cabin altitude. To cause the cabin to descend, the outflow valves are moved toward the closed position to allow less air to escape from the cabin. If a cabin climb is required, the valves are moved toward a more open position to allow more air to escape from the cabin, thereby reducing the pressure differential between the cabin and the outside atmosphere.

The normal mode of operation for most modern pressurization systems is automatic. In a typical system operating in the automatic mode, the pressurization control panel (Fig. 5-2) is used to preset two altitudes into the pressure controller: cruise altitude and destination airport altitude. Takeoff-airport altitude is fed into the controller by the air-data computer at all times when the aircraft is on the ground.

A squat switch on the main landing gear signals whether the aircraft is on the ground or in the air. On the ground, the FLT/GRD switch on the control panel is used to keep the cabin depressurized by driving the main outflow valve full open

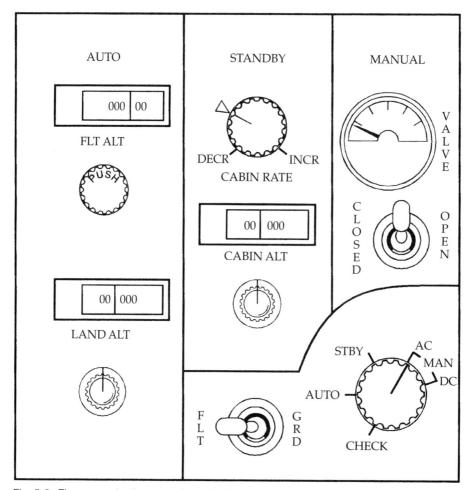

Fig. 5-2. The pressurization control panel allows the crew to control the cabin pressurization in the automatic, standby, and manual modes.

when the switch is in the GRD position. In the FLT position, the switch causes the controller to modulate the main outflow valve toward the closed position, pressurizing the cabin to 0.1 PSI (approximately 200 feet below field elevation). This ground pressurization makes the transition to full pressurization in flight much smoother and dampens the effect of ground-effect-induced pressure changes encountered during the takeoff roll.

In the air, the pressurization controller maintains a proportional pressure differential between the aircraft altitude and the cabin altitude. The controller senses only PSI. The actual PSI of the cabin is translated into approximate cabin altitude. The actual altitude will, of course, depend upon local pressure fluctuations. By "climbing" the cabin proportionally to airplane climb (actually by allowing the pressure of the air in the cabin to decrease), the cabin-altitude rate of change is held to the minimum required in order to provide the most comfortable transition of the passengers and crew. The maximum cabin rate of climb in the AUTO mode is 500 sea-level feet per minute. The maximum rate of descent is 350 sea-level feet per minute.

Approximately 1000 feet below the preselected flight altitude (actually, when the outside air pressure is within 0.25 PSI of the pressure corresponding to the preselected FLT ALT altitude), the cruise relay will trip, scheduling the controller to begin maintaining an isobaric (constant pressure) 7.80 PSI differential between flight and cabin altitudes. During cruise, minor airplane excursions from flight altitude might cause the pressure differential to reach 7.90 PSI to maintain a constant cabin altitude.

At the beginning of the descent, approximately 1000 feet (0.25 PSI differential pressure) below cruise altitude, a descent relay will trip and schedule the cabin to begin a proportional descent to the landing-field altitude set on the LAND ALT dial. The controller will manage the descent so that the aircraft lands slightly (0.1 PSI differential pressure) pressurized so that the rapid changes in altitude and rate of descent often encountered during the approach and landing phase will result in minimum cabin pressure changes.

During taxi, the FLT/GRD switch is repositioned to the GRD position, which allows the controller to slowly move the main outflow valve to the fully open position, fully depressurizing the aircraft. The fully open position of the main outflow valve on the ground also prevents the equipment fan from depressurizing the aircraft to altitudes below field elevation. Figure 5-3 shows the aircraft and cabin-altitude profile for a typical aircraft with the pressurization operating in the automatic mode.

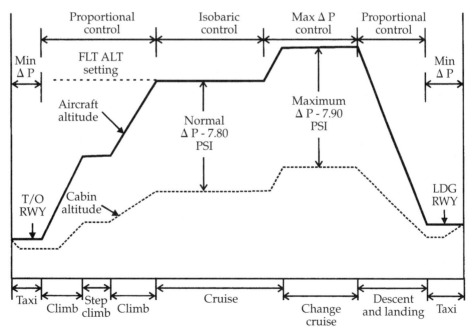

Fig. 5-3. The cabin altitude and the aircraft altitude will vary according to the maximum pressurization differential and the proportional differential during the flight.

If this pressurization system must be operated in the standby mode, the GRD position of the FLT/GRD switch will still drive the main outflow valve to the fully open position. In the FLT position, however, the operation is different from that of the automatic mode. The CAB ALT counter must be set to 200 feet below takeoff-field elevation to pressurize the aircraft properly when the switch is placed in the FLT position prior to takeoff.

Once airborne, the CAB ALT must be set to the isobaric cabin altitude based on the proposed cruise flight altitude and a pressure differential of 7.8 PSI. A cockpit placard contains the information necessary to match the proper cabin and flight altitudes. A proportional rate of climb is set with the cabin rate selector. The minimum selectable cabin rate is 50 feet per minute at the DECR point on the selector. The INCR point will select the maximum rate of change of 2000 per minute. The triangular index on the selector indicates a cabin rate of change of approximately 300 feet per minute.

During the descent, the CAB ALT is set to 200 feet below landing-field elevation to ensure a pressurized cabin during approach, descent, and landing. A proportional rate of descent is established with the cabin-rate selector.

Operation in the manual mode assumes failure of both the automatic and standby modes. MAN AC allows the pilot to reposition the main outflow valve with the outflow valve switch and the ac electric motor. MAN DC allows repositioning using the dc electric motor. The position of the valve is monitored on the outflow valve indicator. The rate of reposition of the valve with ac power is faster than when using dc power. Precise and smooth control of cabin pressurization is extremely difficult in the manual mode.

Figure 5-3 is a graphic representation of the flight and cabin altitudes through the various phases of flight described with this typical system. It should be noted in particular that some aircraft are designed to be landed with the cabin pressurized, as in this system, and other aircraft flight manuals specifically prohibit pressurized landings. Aircraft not stressed for pressurized landings are typically equipped with a method for immediately depressurizing the aircraft upon touchdown. This type of system might not have a facility for allowing the pilot to indicate the landing-field elevation prior to takeoff.

The pressurization control system found on Cessna Citation 500 series aircraft uses a variable isobaric controller to control two identical outflow valves. The outflow valves in this system are spring-loaded to the closed position and are moved toward the open position by regulated vacuum. This system, like the previously described system, assumes a constant inflow of air and modulates that which is allowed to escape from the cabin.

Prior to takeoff, the desired cruise altitude plus 1000 feet is selected on the aircraft pressurization controller dial labeled ACFT. The cabin altitude that will result from maximum pressurization differential at this cruise altitude is displayed on the adjacent scale on the selector marked CABIN. During the takeoff roll, advancement of the throttles operates three solenoid valves to move the outflow valves, which have been held in the full open position until this point, into the pressurization range and pressurize the cabin prior to liftoff to prevent a pressurization spike or "bump" when the main landing gear squat switch goes to the air mode.

By selecting an altitude that is 1000 feet above the planned cruise altitude, the aircraft will never reach maximum differential pressure. This technique eliminates the continual pressure spikes that result from operation at maximum differential. At maximum differential, the outflow valves are at times totally closed. When they are again opened to release excess pressure, a spike, or rapid small change in the cabin pressure, will occur. This is annoying and uncomfortable for passengers. Operation just below maximum differential allows the outflow valves to remain slightly open at all times and so eliminate these spikes.

In preparation for landing, 200 feet above landing-field pressure altitude is set in the controller using the CABIN scale for reference. As the aircraft descends through 200 feet above touchdown, the cabin should be depressurized and the pressurization differential will read 0 PSI. At touchdown, with the throttles at less than the 80 percent N_2 position, the left main landing gear squat switch will open a solenoid valve and route vacuum to the outflow valves opening them fully. With the airplane previously depressurized, this will be unnoticeable.

The outflow valves are held closed by spring pressure in this system and opened by regulated vacuum. In the event of a vacuum failure, the outflow valves will go to the full closed position, and the cabin will go to maximum differential pressure. This might not be noticed by the crew at maximum-altitude cruise, but when a descent is started, the cabin will descend with the aircraft to maintain maximum pressurization differential.

In this event, since controlling the air that is flowing out of the cabin is no longer possible, the source of pressurization air must be eliminated to depressurize the aircraft. This is accomplished by turning the source selector to the OFF position (Fig. 5-4). In the event it becomes necessary to dump the cabin, the manual dump valve can be opened to allow unregulated vacuum to be routed directly to the dome of the outflow valves, moving them to the full-open position and allowing the cabin to climb to the aircraft altitude.

All pressurized aircraft must be equipped with a cabin-altitude warning system to warn the crew if the cabin altitude exceeds 10,000 feet. Pressurization problems are normally detected by the crew through the appearance of pressurization "spikes" or sudden changes in cabin pressure or rate of change. These spikes can be early signs of impending pressurization controller failure or might be the result of dirty and sticking outflow valves.[3] The cabin-altitude warning system is of most benefit in alerting the crew of a missetting in the pressurization controls. A mistake of this type would allow the cabin to smoothly exceed the 10,000-foot limit without the presence of the spikes that alert the crew. Of course, as explained in chapter 3, the crew suffering from hypoxia might not notice this mistake without a visible or audible warning.

Cabin-altitude limit valves

Pressurized aircraft certified to fly above 25,000 feet must have a means to limit the cabin altitude to 15,000 feet in the event of any reasonably probable malfunction or failure of the pressurization system. This can take the form of warning lights to

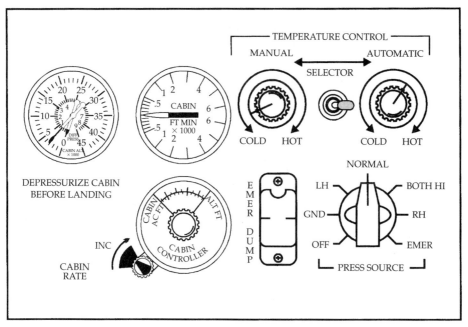

Fig. 5-4. The environmental control panel found on the Cessna Citation aircraft allows the crew to control pressurization and cabin temperature.

alert the crew that action must be taken to limit the cabin altitude or the cabin limiting function can be totally automatic.

In executive-size aircraft, the outflow valves are often pneumatically operated. As described earlier, they are spring-loaded to the closed position and are moved toward the open position by regulated vacuum. In this type system, a failure of the pressurization controller can cause the outflow valves to be moved to the full-open position by providing too much vacuum to the dome of the valve. To limit the cabin altitude in this system, the cabin-altitude limit valve opens to allow cabin air to enter the vacuum lines. This will modify the vacuum applied to the dome of the outflow valves and will allow them to move toward the closed position, consequently limiting the cabin altitude.

RAPID OR EXPLOSIVE DECOMPRESSION

As previously stated, FAR 25 requires that any pressurized aircraft authorized for flight above 25,000 must have a means to limit the cabin altitude to 15,000 feet in the event of any reasonably probable malfunction or failure of the pressurization system. This refers to failures in the system for controlling cabin pressure, not the method of providing or containing it. Explosive and rapid decompressions can be caused by a number of events:

- Failure of the aircraft structure such as occurred in the Hawaiian 737 incident

- Failure of a door or the locking mechanism on a door such as those that occurred on the DC-10 aircraft or the United 747 that lost a cargo door over the Pacific[4]

- Failure or leaks in door seals and other seals between the pressure vessel and the unpressurized areas of the fuselage

- Failure of the air supply into the cabin. Ruptures in ducting from air-conditioning packs or total separation of ducting have been known to happen.

No matter what the cause, an explosive or rapid decompression is the loss of cabin pressure resulting in the rapid ascent of the cabin altitude to the altitude of the aircraft. The rate of this change will depend upon the pressurization differential before the event, the size of the opening through which the air is escaping, and the rate at which air is being provided to the cabin.

Explosive or rapid decompression is one of the few times it is necessary to react rapidly to an emergency in a jet aircraft. It is of utmost importance that the crewmembers immediately don oxygen masks and establish 100 percent or emergency (pressure) flow as appropriate. Checklists to be followed in the event of an explosive decompression will, of course, be dictated by the aircraft flight manual, but many will follow this general flow:

- Oxygen masks   DON, 100 percent
- Crew communications   ESTABLISH
- Passenger masks   DROP
- Emergency descent   ESTABLISH

Emergency-descent procedures will depend upon the structural integrity of the airframe and the manufacturer recommendations, but will follow this general pattern:

- Throttles   IDLE
- Speedbrakes   DEPLOY
- On some aircraft—landing gear   DOWN
- Ignition   ON
- Speed M_{MO}/V_{MO}
- Transponder   7700
- Seat belt light   ON
- ATC   NOTIFY

Whether or not to bank the aircraft and whether or not to extend the landing gear will depend upon the aircraft flight manual. Some manufacturers suggest that if the landing gear has been extended in an emergency descent, the gear be left extended for the remainder of the flight due to possible damage to gear doors.

No matter what procedures the manufacturer recommends, however, it is vital that the crew don the oxygen masks and get the airplane down to a breathable altitude as rapidly as possible. This maneuver will not be the orderly drill that is accomplished during training. There might very well be a very high noise level in the cockpit along with a thick fog and swirling papers and dust. Chapter 3 outlined some of the effects the crew and passenger might suffer as a result of a rapid decompression at altitude. Suffice it to say, however, that it will be a tense situation at best.

Oxygen

Supplemental oxygen must be available in the event of a cabin pressurization failure. For aircraft certified at or above 25,000 feet, a 10-minute supply of supplemental oxygen must be available for each occupant of the aircraft. Depending upon the size of the aircraft, this oxygen might be supplied in one or more bottles or might, in some cases, be supplied partially by bottles and partially by oxygen-generating systems.

Most corporate jet aircraft are equipped with one or more bottles supplying the crew and passenger oxygen systems. In systems that incorporate only one oxygen bottle, a valve will be positioned in the system to isolate the crew and passenger systems. There will also be a provision for automatically dropping the passenger masks when the cabin altitude reaches approximately 14,000 feet and a manual valve to drop masks if the automatic system malfunctions or if oxygen is needed before the cabin altitude reaches 14,000.

A typical cockpit preflight includes ensuring the crew's oxygen masks are plugged in and providing oxygen flow. The oxygen pressure will be checked, and the valve that isolates the crew and passenger systems must be opened to allow deployment of the passenger masks if necessary. Each crewmember should check and adjust the oxygen mask and make sure that it is properly stowed in its quick-don position.

A diagram representative of an oxygen system found in many corporate jets is shown in Fig. 5-5. This system incorporates quick-donning crew masks that provide diluter-demand oxygen, 100-percent oxygen, or emergency oxygen that delivers 100-percent oxygen to the crew mask at slightly positive pressure for high altitudes or respiratory protection against smoke and fumes. The crew oxygen masks incorporate a microphone that is activated by a switch on the side panel that switches communications from the normal microphone to the oxygen-mask microphones.

Passenger oxygen masks are located in compartments in the headliner of the aft cabin. With the passenger oxygen (called the crew-only valve in some installations) in the open position, the passenger masks can be dropped either by action of the automatic aneroid sensing system or by activating the manual drop switch. The aneroid system senses cabin altitudes above approximately 14,000 feet and uses normal dc electric to reposition the automatic valve and allow regulated oxygen pressure to the passenger mask system. It is regulated oxygen pressure that releases the latches on the compartment doors and allows the masks to deploy. As

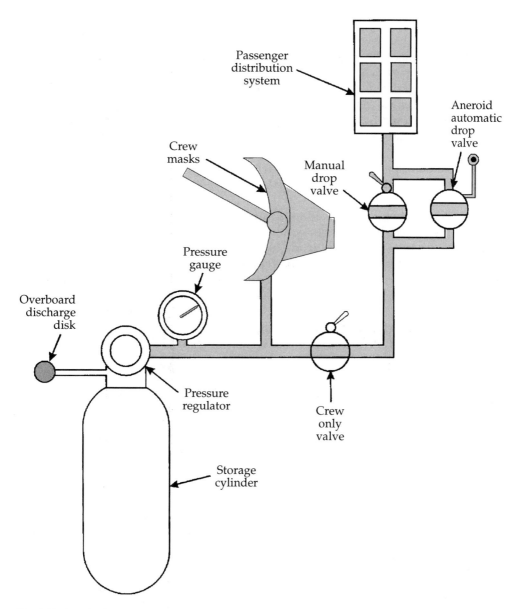

Fig. 5-5. A typical oxygen system will allow the crew to restrict oxygen distribution only to the pilots or also to the passenger cabin; the system will also provide for automatic and manual deployment of the passenger masks.

the passenger pulls down on the mask to don it, an attached lanyard pulls a pin from the supply valve that releases the flow of oxygen into the mask. Passenger masks are typically constant-flow rebreather types. This type mask has a bag attached to the mask. When the passenger inhales, 100-percent oxygen is delivered to the mask cup. Breath is then exhaled into the rebreather bag to conserve the oxygen normally exhaled with each breath.

Manual drop of the passenger oxygen masks is accomplished by activation of the manual-drop switch. This repositions the manual valve to the open position, which routes regulated oxygen pressure to the passenger distribution system. The passenger oxygen valve must be open for the masks to be manually dropped.

The pilot should be aware that unpressurized flight above 30,000 feet can result in incapacitation and unconsciousness even with the use of supplemental oxygen (see chapter 3). If a pressurization problem occurs at these altitudes, it is imperative that an emergency descent be immediately begun. This is not the time to request a lower altitude from ATC. Begin the descent, and then inform the controller.

Passengers should be briefed to breathe as normally as possible when using oxygen. Hyperventilation is common when using oxygen because of the anxiety involved with the incident and tends to perpetuate itself. The passengers should also be informed that the rebreather bag attached to the oxygen mask will not fully inflate. If they have never used passenger masks aboard an aircraft (and most have not), they might expect the bags to inflate and will be concerned that the system is malfunctioning when they do not inflate. This should all take place in the preflight emergency briefing; the midst of an emergency descent after an explosive decompression is not the time to be explaining systems to panicked passengers.

ENVIRONMENTAL SMOKE OR CONTAMINATION

As noted in FAR 25, a means must be available for rapidly equalizing the inside and outside cabin pressures. This might be necessitated by an overpressurization malfunction, but it might also come about as a means of eliminating smoke or other contamination from the cabin air.

Cabin air contamination might come from various sources. Anything that contaminates the air moving through the compressor will eventually show up in the cabin air supply. Oil in the compressor case is not only bad news for the longevity of the engine, it might fill the cabin with smoke. Other sources of cabin air contamination are fires in or around the ducting, electrical fires, bird ingestion, and the like.

The first item on every contamination and smoke removal checklist is OXYGEN, DON & 100 percent. The crew needs to breathe to solve any problems. Electrical fires and smoke are usually distinctive for their burning-wire smell and will be handled by electrical systems checklists. Environmental smoke and odor may be harder to track down. For aircraft having more than one air-conditioning pack, the air from each pack can be shut down in order to see if the contamination is coming from that pack. Of course, sufficient time must elapse for the air to be changed in the cabin to know if the proper source has been isolated. A corporate jet will usually totally change the air in the cabin every three to seven minutes. A significant lessening of the contaminants should be noticed before this time, however, if the proper source has been isolated and eliminated. For aircraft with only one air-cycle machine, the source of bleed air can be changed from one engine to the other to make a similar determination.

Smoke or contamination removal does not usually require that the cabin be dumped, but if the smoke contamination is severe and cannot be isolated and eliminated, the cabin must be evacuated. This will be an extreme solution, and other solutions, as dictated by the aircraft checklists, should be tried first. If all else fails, however, the passengers should be seated and belted, the masks dropped in the passenger cabin, and the emergency cabin-dump valves opened. This will cause a rapid evacuation of the cabin, and of course will probably necessitate an emergency descent.

AIR CONDITIONING

Most new jet pilots believe that the air must be warmed before it enters the cabin and are shocked to find that just the opposite is true. The source of pressurization air is the compressor case of the engine, and the air bled from this source is anything but cold. The temperature of air entering the air-cycle machine, or air-conditioning packs, can be as hot as 500°F. Even at altitude, the air must be cooled before it enters the cabin.

Air-cycle machine

Corporate aircraft use the term air-cycle machine, or ACM, and airlines use the term air-conditioning packs, or simply packs, to describe what is the same machine. Whatever the terminology used, the machine has only one function—it cools air. The air-cycle machine is the source of cabin pressurization air. One such machine is usually sufficient for corporate aircraft. Airliners usually have two or more.

Air is taken from the compressor section of the engine as early in the compression process as is possible to obtain a sufficient air supply. The earlier the air is extracted from the compressor, the less that compressor efficiency is impacted. Most bleed-air extraction schemes will take air from an intermediate compressor stage[5] for most normal operations and supplement this air with air from a later stage when the bleed air extracted from the earlier stage is insufficient to meet pressurization needs. This can be due to altitude or due to use of part of the bleed air for anti-icing needs.

Then the hot air can be routed through precoolers in the engine nacelle that use bypass air through an air-to-air heat exchanger to remove some heat energy from the air. A diagram of a typical air-cycle machine is shown in Fig. 5-6. The air from the engines is routed to the first heat exchanger, which reduces the temperature of the air before it is routed to the compressor. The compressor then increases the pressure of the air, which also increases the temperature. The compressed air is then routed through another heat exchanger that cools it somewhat before it is routed to the turbine. Part of the internal energy of the air is used to rotate the turbine. The rest of the heat energy is lost in the expansion chamber.

When I tell most new jet pilots that the air-cycle machine takes hot air from the compressor case, compresses it more, and makes it hotter to cool the airplane, they look at me as though I just landed from another planet. Try this experiment. Hold

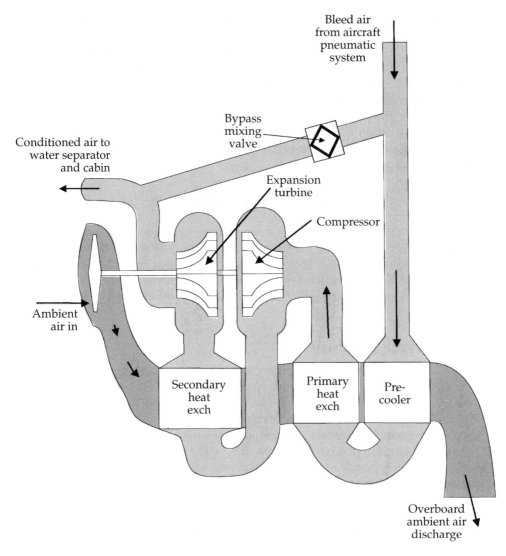

Fig. 5-6. The air-cycle machine, also known as an air-conditioning pack, cools the bleed air prior to its introduction to the cabin. The mixing valve adds hot bleedair to the cooled air to provide the temperature commanded by the crew.

your hand in front of your mouth, and, with your mouth wide open, blow on your hand as though you were going to clean a pair of eyeglasses. The air that hit your hand felt warm, didn't it? Now, pucker your lips and blow on your hand again. The air that hit your hand felt much cooler didn't it? Well, you just made a miniature air-cycle machine. Obviously the temperature of the air from your lungs remained constant. In the first test, the air exited your mouth at approximately the same volume as it left your lungs. In the second test, the air was forced to go through a constricted duct, your pursed lips, and so the air must expand as it leaves your lips. This is *adiabatic expansion*, and this expansion of the air cools it.[6] That is exactly how air-cycle machines and air-conditioning packs work.

Temperature control

The only thing that the air-cycle machine is able to do is make cold air. Obviously, at FL370 we will need less cooling of the air provided to the cabin than we will need on the ground at Las Vegas. Many different temperature-control systems are used to provide just the right temperature to the cabin, but they all work under the same principles. A hypothetical automatic temperature-control system will consist of a network of temperature sensors installed in the cabin and in the air delivery ducts directly downstream from the air-cycle machine.

These sensors deliver their input to a temperature controller that also receives inputs from the temperature selector switch located in the cockpit. The temperature controller will integrate these signals and will send a signal to the bleed-air mixing valve to vary the proportion of bleed air routed to the air-cycle machine to the bleed air mixed into the cooled air to control the temperature. A schematic of a hypothetical system is shown in Fig. 5-7.

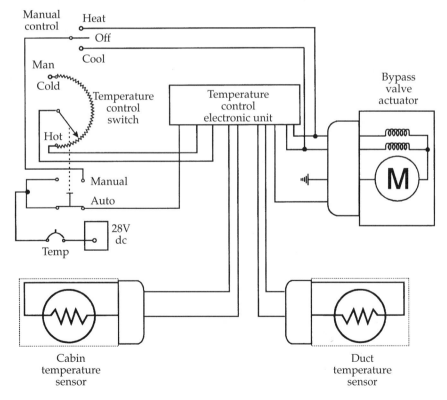

Fig. 5-7. A temperature-control system will take inputs from the passenger cabin-temperature sensor, the duct-temperature sensor, and the cabin-temperature rheostat to compute the position of the mixing valve.

The temperature-control system might be electronic or pneumatic, but most will have both an automatic and a manual mode. The automatic mode will use the integration of the signals from all temperature sensors to automatically control the temperature of the air entering the cockpit. The manual mode will simply send a signal to the bleed-air mixing valve to manually reposition it to mix more or less hot bleed air into the cooled air.

Air-cycle machines and air-conditioning packs incorporate a water separator to remove humidity from the air prior to routing it to the cabin. Removal of humidity is not usually necessary at cruise altitudes because the air at altitude contains virtually no moisture, but during ground operations or low-altitude maneuvering, especially in humid climates, the water separator is a necessity. Most water separators consist of a swirl chamber and a condenser sock to cause any water vapor present to condense and so be removed from the airflow. This is mostly of use at low altitudes.

Air at the flight levels is extremely dry. The very cold air found in the flight levels just is not capable of holding water, and the cabin of a jet at altitude usually has about 5 percent relative humidity. Most deserts aren't that dry. Flying long hours in such a dry environment will cause dehydration, even in the absence of the other dehydrating agents (coffee) often ingested by pilots. Dehydration will add to your fatigue level and make you much less tolerant to hypoxia. Drink lots of water or fruit juices during flight to help ward off dehydration and its ancillary effects.

Air distribution

Air is distributed from the air-conditioning packs or air-cycle machine(s) to the passenger cabin and cockpit through air distribution ducts. Normally the air supplied to the overhead (gasper) air outlets will be cooler than that supplied through the floor and sidewall ducting. If this air is not from a separate supply, the overhead ducting system will normally be blocked above a preset distribution air temperature. Warm air blowing into passengers' faces can increase any tendency to air sickness and therefore is best restricted. Various arrangements of automatic and manually controlled fans may be installed to aid in air circulation.

Warnings and malfunctions

The most common cause of air-cycle machine failure or shutdown or "pack trip" is an overheat condition. This is caused by demanding too cold a temperature from the air-conditioning system. When an excessively cold temperature is demanded, the air-cycle machine works on a larger supply of air. Notice that the first stage of the air-cycle machine is a compressor that compresses the air received from the engine supply. The more the air is compressed, the more heat there will be. Eventually, this reaches a limiting temperature, and the pack "trips" or shuts down to cool off.

In an airliner, when only one pack trips, any other pack or packs provide sufficiently conditioned air and maintain cabin pressurization. In a smaller aircraft with only one air-cycle machine, an alternate means of providing emergency pressurization must be provided. This might take different forms but usually is provided by means of bleed air directly from the engine or engines. This might be slightly cooled in a mixing chamber but is usually quite hot and quite noisy.

The remedy for an air-cycle machine shutdown or a pack trip is to reduce the demand upon the system by commanding a warmer temperature. Refer again to Fig. 5-6. You will notice that when a warmer temperature is commanded, the bleed-air mixing valve is opened. This has two effects; it mixes hot bleed air with the cooled air from the air-cycle machine to modify its temperature and, more importantly, it takes this air from the bleed-air line prior to that air being routed to the compressor and turbine of the air-cycle machine. With less air to work on, the machine cannot overheat.

If the temperature commanded is too warm, a duct overheat light will be noticed. This signals that the air in the duct leading to the passenger cabin is too warm. The remedy for this problem is the obvious one: Turn down the temperature. Most problems with duct overheat and pack trip will not be encountered with a properly functioning automatic temperature control system.

FACTORS AFFECTING AIR-CONDITIONING EFFICIENCY

The air-cycle machine or air-conditioning packs work by compressing and expanding air. Anything that takes the place of air volume, such as humidity, will decrease the efficiency and the cooling capability of an air-cycle machine. Because the air-cycle machine works by compressing and expanding a parcel of air, any water present in the air as humidity will decrease the volume of air that the air-cycle machine has available to cool. Water is considered an incompressible fluid; therefore, the higher the relative humidity, the less cooling of the air that can take place.

Another factor that affects the apparent efficiency of the air-cycle machine is heat soaking of the aircraft. As the airplane sits on the ground in a high ambient temperature, the airplane and the ducting that carries the conditioned air becomes heat-soaked. As cooled air is routed from the air-cycle machine to the passenger cabin, it will gain heat from these heat-soaked parts. The ducting must be cooled significantly before noticeable cooling of the cabin can take place. For this reason, some smaller aircraft have a method of bypassing the heat-soaked ducting in the aircraft body by routing air directly from the air-cycle machine to the rear of the passenger cabin through larger ducting. This cooling air can be termed flood cooling or auxiliary cooling, but its use is normally limited to low altitudes to ensure that an adequate supply of pressurization air is distributed through the normal ducting.

Freon air conditioning

Freon air conditioning is installed on many corporate aircraft to augment the cooling provided by the air-cycle machine on the ground and at low altitudes. Freon air-conditioning systems are very similar to those found on automobiles. The system components are conventional with a compressor driven by a small electric motor driven from one of the aircraft generators. Air is drawn through the evaporator and circulated through either the normal aircraft air distribution ducts or through special ducting, depending upon the installation.

Use of Freon air in most installations is limited to lower altitudes (below 18,000 feet) because of the limitations of the compressor motor. At higher altitudes, the compressor motor makes an excessive demand upon the aircraft electrical system. Also, Freon is prohibited from use with only one aircraft generator on line or with other high loads on the electrical system.

Freon air can, however, prevent the aircraft from becoming heat soaked when sitting on the ramp in high ambient temperatures. If the aircraft can be connected to an external power unit, the Freon air conditioner can be operated on the ground without the engines running to keep the cabin cool and provide more comfort for passengers during ground operations.

Endnotes

1. The FAA and I continue to disagree as to just what proper oxygenation of the blood ought to be for pilots, but thankfully each pilot can control his or her own blood oxygen levels given the proper equipment.

2. This expansion and contraction over wide temperature ranges causes metal fatigue that can, if not detected and corrected, cause catastrophic airframe failure. Who among us can forget the awesome sight of the Aloha Airlines 737 with its cabin roof missing due to just this type of fatigue.

3. Sticking outflow valves used to be almost exclusively caused by a coating of nicotine from cigarette smoke flowing out of the cabin. Many corporate flight departments banned smoking aboard their aircraft prior to the FAA action to ban smoking on airline flights. Outflow valves do not get gummed up nearly as quickly as they used to. Instrument air filters also last longer before requiring cleaning or change.

4. Dave Cronin, the captain of that flight, readily admits that if the aircraft had attained a higher altitude prior to the failure, the outcome might have been far more disastrous. The failure of the cargo door occurred in the low 20,000s. Refer to chapter 3 to review the changes that occur in the oxygen dissociation curve in the middle 20,000s.

5. The Boeing 737-200 extracts cabin pressurization bleed air from the 8th stage of the compressor for all normal operations and supplements this with 13th stage bleed air when the 8th-stage air is insufficient.

6. The process of expanding or compressing a gas so that no heat enters or leaves a system is said to be adiabatic, from the Greek "impassible." Adiabatic process can be achieved either by compressing or expanding a gas so rapidly that heat has little time to enter or leave. This is in accordance with the first law of thermodynamics, which is simply stated as the thermal version of the law of conservation of energy.

6

Miscellaneous systems

NO EFFORT WILL BE MADE TO COVER ALL POSSIBLE SYSTEMS FOUND IN jet aircraft. After all, fuel systems might have many tanks and many pumps, but they are, after all, just fuel systems and all pilots have been dealing with such systems since their first hour of dual. It also can be said that some of the fuel systems found on general aviation piston twins are far more complex than those found on some jets. At any rate, the assumption will be made that the reader has a working familiarity with the types of systems that are common in the general aviation fleet, and we will concentrate our discussions on those that are peculiar to jet aircraft, or at least large transport aircraft.

ANTISKID AND BRAKING SYSTEMS

Power brakes are found on most automobiles, but that has not always been the case. If you have ever driven a car without power brakes, you will remember the greater pressure on the brake pedal required to stop the car. It would seem, from listening to automobile commercials, that power brakes and antilock braking systems (ABS) are a brand new technology developed in Detroit solely for the safety of the motoring public. This technology has been available on aircraft for almost 40 years, however. As with any other system, there are many different variations of power brakes and antiskid or antilock braking systems, but regardless of the differences, they help stop the aircraft and remove braking pressure from a wheel that is skidding.

Power brakes

Jets are typically equipped with power disc brakes that are operated hydraulically. Some systems use the main aircraft hydraulic system; others are separate systems. The basics of a power-braking system include the master cylinders attached to each pair of rudder pedals—the cylinders increase the pressure applied by the pilot to the brake pedals—and the stationary and rotating components of the brakes themselves. Aircraft braking is differential; it is possible to apply different amounts of pressure to the brakes on each side of the aircraft. This obviously helps in turning the aircraft in tight places.

Figure 6-1 shows the power-braking system found in the Cessna Citation 500 series aircraft. With the gear handle in the DOWN position, dc electric power is applied to an electric pump found in the nose compartment of the airplane. This pump provides pressure to the brake hydraulic system. This provides pressure to multiply the pressure put on the hydraulic fluid by the master cylinders by a factor of two. The power-brake system in this aircraft is necessary to the operation of the antiskid system.

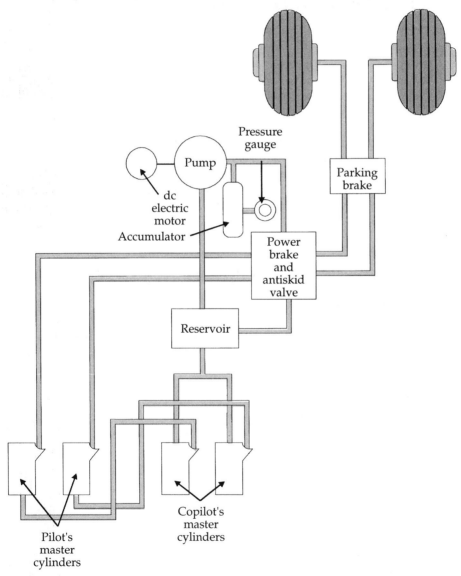

Fig. 6-1. This power-braking system used a dc electric motor to work a pump providing hydraulic pressure to double the amount of pressure applied to the master cylinders by the pilot and route it to the wheel brakes.

Skids—what and why

Skids occur when the wheels are not rolling on the runway. Sounds simple, doesn't it, but that is what is happening. When brake pressure is applied too heavily, or when the surface on which the wheel is rolling is slick, the wheels might stop turning, or lock, and just slide along the runway. It might seem that would create quite a dragging force and decelerate the aircraft, but it only puts skid marks on the runway.

Braking is accomplished by a controlled deceleration of the aircraft wheels. This causes the whole aircraft to decelerate, but the main emphasis is on the wheel. The brakes, through application of hydraulic pressure, force the brake pads against the discs and create a frictional force to reduce wheel rotation. The frictional force generated between the tires and the runway causes the airplane to decelerate and stop.

As a pilot applies brakes on a dry runway, the friction rises until an 18–20 percent rolling skid is reached. At this point, the wheels are still rolling at 80 percent of their normal speed, and maximum friction is generated at the tire-runway interface. Almost no slipping or skidding of the tire has actually occurred. When this point of peak friction has passed, however, braking friction diminishes rapidly as the wheel locks in a total skid. When tires begin to skid, the wheels are no longer turning, and the brakes are no longer absorbing any of the energy. The tires on a locked or skidding wheel will be the recipient of all the deceleration energy with extremely unfavorable effects on tread life. Skidding tires are uncontrollable. When the tire wears through, it blows out, and the airplane becomes a handful.

Modern jet aircraft tires are equipped with high-speed and high-pressure tires. When one of these tires begins to skid, small pieces of rubber are torn off and become like small rollers beneath the tires. Next, the heat generated by the skid begins to melt these small pieces of rubber into an extremely effective lubricant beneath the tire, totally negating any traction that might have been available. The aircraft will actually turn away from the skidding wheel. Brakes are now totally ineffective, and the only way to regain control of the situation is to totally release the brakes to get the wheels rolling again. This all happens so quickly, however, that the tires are usually blown first.

It is extremely difficult to judge the point at which the 18–20 percent skidding force has been exceeded and the aircraft is beginning a full skid until the situation is well developed. The tendency, without antiskid, is to apply too little brake pressure in an effort to avoid the skid and so greatly increase the landing roll.

Early devices were skid-warning systems. A transducer in the axles sensed a differential in the wheel speeds or an excessive deceleration of one or more wheels and transmitted a signal to the pilot who then was required to release brake pressure.[1] The pilot was then required to reapply brakes until the release signal was again received. The result was an uneven and prolonged deceleration. The optimum, of course, was to find that point just shy of the release point and hold the brakes there.

The earliest true antiskid systems were unmodulated. These systems simply compared the rate of wheel-speed change with a predetermined rate standard. When this standard was exceeded, brake pressure was automatically released.

This system only prevented total wheel lockup. A major drawback with this system was that no pressure was applied to the brakes during the release phase. This, of course, prolonged the landing roll. During heavy emergency braking, as might occur during a rejected takeoff, heavy vibration or bumping could occur as the brake-release valve rapidly shuttled on and off.

Modern antiskid systems are fully modulated or proportional. These units are sensitive to the rate of wheel-speed change, but improvements in electronics and faster-acting valves allow the system to remove and replace brake pressure several times per second if necessary. These systems allow automatic reapplication of brakes and so are able to keep the brakes just shy of the skid point in the area of optimal braking.

Antiskid components and operation

Figure 6-2 is a diagram of an antiskid system found on the Cessna Citation series aircraft. With the landing gear handle in the DOWN position and the antiskid switch ON, the power-brake/antiskid system receives dc electric power from the aircraft's main electrical system. Transducers in each main landing gear wheel transmit a speed signal to the electronic control box. Detection of sudden deceleration in a wheel (impending skid) causes a signal to be sent to the antiskid valve to reduce pressure being applied to the brakes.

When the transducer signal indicates that the wheel has returned to normal speed, the brake pressure is reapplied. The system requires that the wheels be spun up to a speed equivalent to a ground speed of 12 knots before being activated. This prevents landing with locked wheels.

With a fixed amount of brake pressure, the rate at which a wheel slows down is dependent upon how much force the tire can exert against the runway surface before skidding. For example, on an icy runway, the tire can exert very little force against the runway before it skids. Without antiskid, the wheel stops almost immediately and begins to slide. Because the antiskid system senses the rate of deceleration of the wheel, it in effect senses the coefficient of friction of the runway surface. By modulating and controlling the pressure to the brakes, the antiskid system can give the maximum allowable braking effort for the condition of the runway. The antiskid valve can only modulate the brake pressure provided by the pilot and can never provide more braking pressure than called for by pilot input.

Proper operation of an antiskid system requires only that the pilot apply firm pressure to the brakes and hold it until the aircraft slows to taxi speed. Pumping the brakes will only defeat the automatic operation of the antiskid system and greatly reduce the effectiveness of the system and greatly increase the runway distance required to stop the airplane. Maximum weight should be on the main landing gear during braking, which would dictate deployment of all lift-dumping and drag devices.

Automatic braking systems

Some aircraft, such as the 737, are equipped with an automatic braking system that works in concert with the antiskid system. This system allows the pilot to preselect

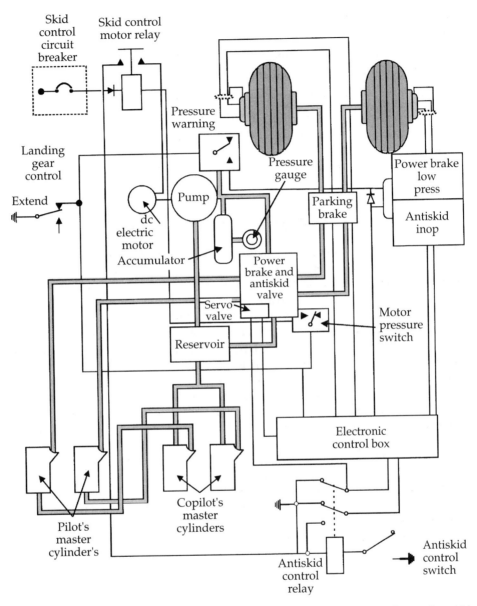

Fig. 6-2. This antiskid system uses transducers on the wheels to sense an impending skid and remove brake pressure from the affected wheel.

a desired acceleration level: minimum, medium, or maximum. The automatic brakes will bring the aircraft to a complete stop unless the pilot terminates the automatic brake operation. Typically, the automatic brakes will be activated whenever the antiskid switch is in the ON position, the thrust levers are retarded, the main wheels are spun up, and the speed brakes are deployed. Automatic brakes are typically deactivated by one of the following actions:

- Depressing one or both brake pedals and releasing them
- Fully depressing one or both brake pedals
- Advancing the throttles (go-around protection)
- Placing the antiskid switch in the OFF position

The aircraft flight manual will contain landing field-length requirements for the use of automatic brakes and antiskid. To convince yourself how effective antiskid and automatic braking systems are, just notice the penalty in additional runway required when these systems are inoperative.

NOSEWHEEL STEERING

Rudder-pedal-actuated nosewheel steering is standard on most general aviation propeller aircraft.[2] But that is not always the case on general aviation jets or airline jets. Most of these types of aircraft employ some means of hydraulically actuated nosewheel steering to replace or to augment rudder-pedal steering.

Nosewheel steering in airliners is usually accomplished by means of a steering wheel or tiller located on the captain's side console. This steering mechanism will allow the nosewheel to be deflected to large angles on either side of centerline. The rudder-pedal steering on these aircraft is considered secondary steering and allows rudder deflections of about 10° either side of center. Rudder-pedal steering is used for takeoff and landing rolls and for taxiing when only small turns are required. These figures, of course, will vary somewhat with aircraft manufacturer, but the basics will remain the same. Nosewheel steering is typically deactivated or disconnected when the nosegear strut is fully extended.

Learjet aircraft have an electrically actuated nosewheel steering system that is variable in authority dependent upon the ground speed of the aircraft. The nosewheel can be turned 45° either side of center at speeds below 10 knots and decreases to 8° at 45 knots. Maximum nosewheel steering speed is 45 knots. The nosewheel steering is activated by a switch on either control wheel or by depressing the STEER LOCK master switch. Steering commands are then given by conventional rudder pedal movement and transmitted through the nosewheel-steering computer-amplifier to the steering actuator. Nosewheel steering is deactivated in the air.

GROUND SHIFT/SQUAT SWITCH FUNCTIONS

Many aircraft systems or functions might be designated for use only on the ground or only when airborne. Other aircraft systems might have warning functions added or removed depending upon whether the aircraft is in the air or on the ground. The method for providing these warnings and/or lockouts is a squat switch (or squat switches) located on either the nosegear or the main landing gear. A squat switch is simply a switch that changes state depending upon whether the gear strut is compressed, indicating ground operation, or at full extension as in flight. These switches are called by various names (ground shift, air/ground safety sensors, etc.), but they all perform essentially the same functions.

Some of the functions typically connected to these squat switches are as follows:

- APU generator switch that can be connected to only one dc bus airborne, both on the ground.
- Landing gear control switch that can be moved while airborne to retract gear and locked in the down position on the ground.
- Stall warning switch that is operative while airborne and inoperative on the ground.
- Flight recorder switch that is fully operative airborne and operates on the ground only when an engine is running.
- Voice recorder switch that prevents erasure airborne; tape can be erased when the parking brake is set.
- Emergency pressurization switch is enabled while airborne and disabled on ground.
- Antiskid switch that while airborne prevents locked-wheel protection on touchdown and on the ground allows wheel spinup and normal operation.
- Thrust reverser switch that prevents deployment airborne but allows deployment on ground.

This list will vary, of course, depending upon manufacturer and model of the particular aircraft, but the functions listed are the most common.

LEADING-EDGE DEVICES

Leading-edge devices are in two basic groups: leading-edge flaps and leading-edge slats. The main difference between the two is that the flaps are hinged and the slats move out forward from the wing to create a slot. Because the swept wing is relatively inefficient at low speeds, a means must be employed to increase the wing area and camber to keep the stall speeds, and consequently the takeoff and landing speeds and field lengths, within reason. Leading-edge devices, in concert with trailing-edge flaps, accomplish this end.

The operation of leading-edge devices is either totally automatic or operation is in conjunction with trailing-edge flaps. Some aircraft, such as the DC-9, provide for operation of the slats independently of the trailing-edge flaps if desired. Slats can be made to deploy and retract in response to changing angle of attack and consequent changing airflow over the forward portion of the wing[3] or can be extended as a function of and in conjunction with trailing-edge flap deployment. Operation of controls in larger aircraft is usually accomplished by use of hydraulic power, and the leading-edge slats and flaps are no exception.

Aircraft with leading-edge devices will also contain a warning system to alert the crew if the leading-edge devices are not positioned in the takeoff position when the throttles are advanced for takeoff.

FLIGHT SPOILERS AND SPEED BRAKES

It is common for jet aircraft to augment aileron control with flight spoilers. Deflection of ailerons activates the flight-spoiler system, which will cause the spoiler(s) on the wing with the up aileron to rise proportional with aileron deflection. The spoilers on the wing with the down aileron will remain faired to the wing. The flight spoilers are normally hydraulically controlled.

Flight spoilers are also sometimes known as speed brakes when they are symmetrically deployed on both wings. Speed brakes are simply high-drag devices extended into the airstream. Technically and aerodynamically there is a difference between speed brakes and spoilers. Speed brakes are drag devices that allow high rates of descent to be used without the consequent buildup in speed. Speed brakes are a secondary flight control and do not rotate the aircraft around any of its three axes. Speed brakes produce drag without affecting lift or without causing the aircraft to pitch.

Spoilers disturb the smooth flow of air over an airfoil and destroy part of the lift of the airfoil. Spoilers cause a pitch change with extension and retraction. High-performance sailplanes typically have spoilers to decrease lift and minimize float during landing.

PNEUMATIC SYSTEMS

The pneumatic system distributes the high-temperature compressed air from the engine (bleed air) to the systems that require bleed air to function. We have already discussed the environmental and pressurization system uses for engine bleed air. Chapter 4 discussed the uses of bleed air within the engine itself and for engine ice protection.

The pneumatic system can provide air for a variety of other uses and systems. Door seals are normally inflated by the pneumatic system. Air-driven gyros are driven by the pneumatic system. Vacuum for operation of pressurization controllers, outflow valves, and emergency dump valves are operated by the pneumatic system through ejector valves. The pneumatic system can also supply crossover, or crossfeed bleed air, from an operating engine for rotating the second engine during engine start.

AUXILIARY POWER UNITS

Most airliners and many of the medium and large corporate jets have auxiliary power units (APUs). These are simply small, self-contained gas-turbine engines, usually located in the tailcone of the airplane. A control panel in the cockpit allows the crew to start and monitor the APU. Gauges monitor APU exhaust gas temperature, RPM, oil pressure and temperature, and electrical output. The APU has its own fire-detection and protection system similar to that found on the main aircraft engines. Fuel is taken from the main aircraft supply and electrical energy for starting is provided by the aircraft batteries.

The APU is used on the ground for generation of electrical power, cooling air, and pneumatics to start the aircraft engines. This makes the aircraft independent of ground-service carts for electrical power and cooling during gate and ramp operations. The APU can also be used airborne as an auxiliary electrical source.

GROUND-PROXIMITY WARNING SYSTEMS

The ground-proximity warning system (GPWS) is designed to provide a warning for five different modes of operation considered as potentially dangerous flight situations. The system that is active between 50 and 2450 feet (radio-altitude height) will provide warnings for the following situations (100 feet for modes 5 and 6):

- Mode 1—Excessive descent rate
- Mode 2—Excessive terrain-closure rate
- Mode 3—Altitude loss after takeoff or go-around
- Mode 4—Unsafe terrain clearance when not in the landing configuration
- Mode 5—Excessive deviation below an ILS glide slope
- Mode 6—Below selected radio-altitude setting

The GPWS system consists of a computer, a PULL UP warning light on each pilot's instrument panel, and a GROUND PROXIMITY fail light. The system also provides vocal warnings over the cockpit speaker. The computer typically receives inputs from the radio altimeter, the air-data computer, the glide-slope receiver, landing-gear lever position, and flap position. The ground-proximity warning system will not warn the crew of flight toward vertically shear terrain or of slow descents to unprepared ground prior to the runway during the landing phase.

Mode 1 has two areas in the excessive descent rate envelope in which it will provide warning. Mode 1 monitors radio altitude and rate of descent. The warnings in this mode are independent of aircraft configuration. If the aircraft penetrates the first area on the descent-rate envelope, an aural warning "SINK RATE" will be repeated each three-quarters of a second on the cabin speaker and the visual PULL UP warning light will illuminate. If the aircraft continues and penetrates the second area of the descent-rate envelope, the aural warning "WHOOP, WHOOP, PULL UP" will be repeated on the cabin speaker until the rate of descent is corrected. Figure 6-3 shows the two areas of alert on the excessive-descent-rate envelope.

Mode 2 of the ground-proximity warning system monitors airspeed, radio altitude and radio-altitude rate of change, barometric altitude rate of change, and airplane configuration. Mode 2 also has two areas in the terrain closure rate envelope. If the aircraft penetrates the first area of the envelope, the aural warning "TERRAIN" is repeated twice followed by the repeated warning "WHOOP, WHOOP, PULL UP" if the aircraft continues into the second area of the envelope. Subsequent operation is dependent upon the aircraft configuration. If the gear and flaps are not in the landing position when leaving the PULL UP area, it will be necessary to gain 300 feet of barometric altitude before the TERRAIN warning message is silenced.

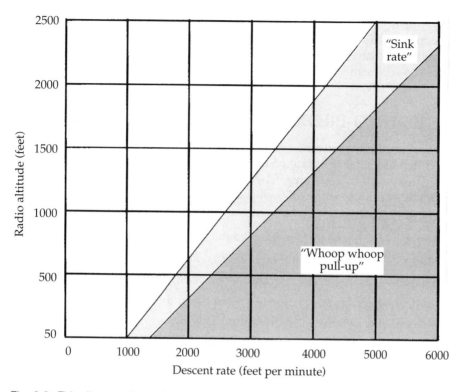

Fig. 6-3. This diagram shows the two areas of the excessive-descent-rate envelopes referenced in Mode 1 of the ground-proximity warning system. (Ineffective below 50 feet.)

As the airspeed increases through M 0.45 with the gear up, the highest radio altitude at which the Mode 2 alerts will be heard is 2450 feet.

When the gear and flaps are in the landing configuration, the lower boundary cutoff will vary between 200 feet and 600 feet, depending upon the barometric rate of change. In this configuration, no altitude gain is necessary to silence the "TERRAIN" warning after the "PULL UP" alert has been sounded. Below 700 feet radio altitude, with the gear and flaps in the landing configuration, the "PULL UP" alert will be disabled and only the "TERRAIN" alert will sound. Figure 6-4 shows the envelope for the Mode 2 warnings.

Mode 3 monitors aircraft configuration, radio altitude and barometric rate of change, and provides an alert if a descent is made during initial climb or during a go-around. Mode 3 is effective between 50 and 700 feet radio altitude and generates a "DON'T SINK" warning repeated every 1.5 seconds when the accumulated barometric altitude loss equals approximately 10 percent of the existing radio altitude. This alert continues until the flight condition is corrected. Mode 3 will also be armed on descent below 200 feet when the aircraft is in the landing configuration. Figure 6-5 shows the Mode 3 warning envelope.

Mode 4 is armed after takeoff and climb through 700 feet of radio altitude. When the first area of the warning envelope is penetrated with an airspeed below

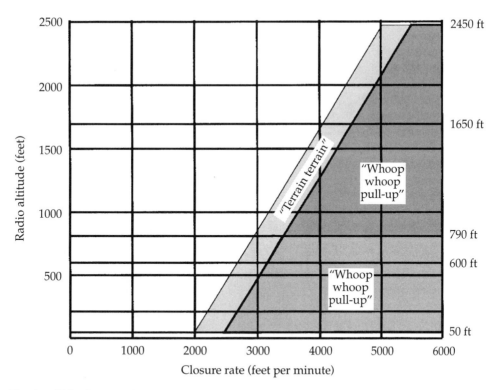

Fig. 6-4. This diagram shows the two areas of excessive descent rate and terrain closure envelopes referenced in Mode 2 of the ground-proximity warning system.

approximately 180 KIAS, the aural alert "TOO LOW GEAR" is sounded. When the airspeed is above 180 knots and the aircraft enters the second area of the warning envelope, the warning "TOO LOW TERRAIN" is sounded, and the upper boundary of the warning envelope is increased to 1000 feet radio altitude. The applicable warning message is repeated every 1.5 seconds until the flight condition is corrected.

Mode 4 also has two warning areas. When the landing gear is down but the flaps are not in the landing position and the airspeed is below approximately 150 KIAS, the warning "TOO LOW FLAPS" will be heard. If the airspeed is more than 150 KIAS, the boundary of the warning area increases to 1000 radio altitude, and the warning "TOO LOW TERRAIN" will be heard every 1.5 seconds until the flight condition has been corrected. The TOO LOW GEAR alert takes priority over the TOO LOW FLAPS warning. Figure 6-6 shows the warning envelope areas covered by this mode.

Mode 5 monitors radio altitude and the captain's glide slope receiver and alerts the flightcrew of a descent of more than 1.3 dots below an ILS glide slope. This warning envelope also has two areas that are accompanied by different warnings. Both warning areas have the warning "GLIDE SLOPE" repeated with increasing frequency as the deviation below the glide slope increases and the radio

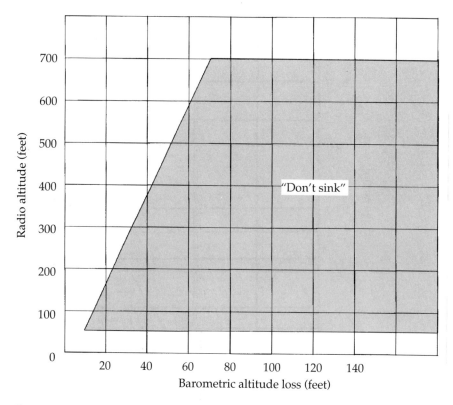

Fig. 6-5. This diagram shows the area of descent rate after takeoff or go-around referenced by Mode 3 of the ground-proximity warning system.

altitude decreases. Below 300 feet radio altitude, the amplitude of the warning message will increase. The warning message can be canceled when in the soft warning area of the envelope but will be rearmed if the aircraft climbs above 1000 feet radio altitude.

Mode 1 through 4 aural alerts have priority over Mode 5, but both the PULL UP lights and the BELOW G/S lights might be illuminated at the same time. Figure 6-7 shows the Mode 5 warning envelope.

Mode 6 monitors the captain's radio altimeter and operates between 50 and 1000 feet radio altitude. When the aircraft descends through the radio altitude set on the captain's radio altimeter, the aural alert "MINIMUMS, MINIMUMS" will sound one time. This mode will be reset if the aircraft climbs above the set altitude. It should be noted that the accuracy of this mode and warning will be degraded for any approach other than a Category II ILS due to the possible uneven nature of terrain prior to the paved surface of the runway.

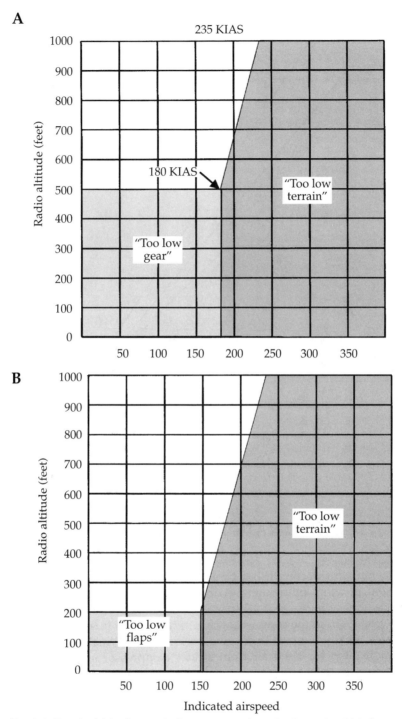

Fig. 6-6. Part A of this diagram indicates areas of terrain closure in which the warning will sound when the landing gear is not extended. Part B shows the areas of terrain closure in which the warning will sound when the flaps are not in the landing position in Mode 4 of the ground-proximity warning system.

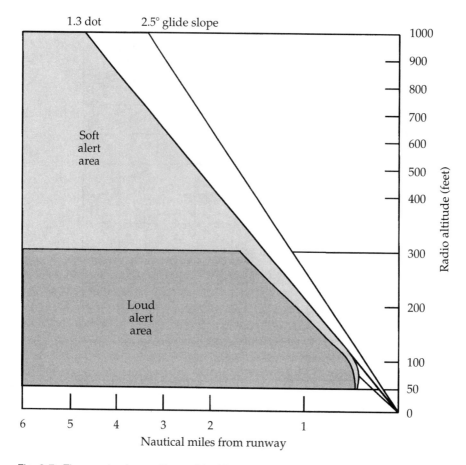

Fig. 6-7. The warning horn will sound in either soft or loud modes depending upon glide-slope deviation and radio altitude in Mode 5 of the ground-proximity warning system.

ICE AND RAIN PROTECTION
Anti-icing

Anti-icing in jet aircraft is accomplished by a combination of bleed air and electrical power. Airfoil, or wing, anti-icing is accomplished by routing engine bleed air to ducts in the wing leading edges and subsequently to the leading-edge devices. This bleed air is taken from the compressor case at the earliest possible stage so as to have the least negative impact on engine performance. When necessary, this bleed air is augmented by bleed air from a later compressor stage. Crossflow valves allow the wings to be anti-iced from either or both engines.

The ground sensor, or ground-shift switch, prevents the wing anti-icing from being used on the ground except for testing because of the excessive temperatures that would result without the cooling airflow available in flight. A thermal switch in the wing leading edge will close the bleed-air valves if the temperature of the leading edge becomes excessively high during ground test. During normal operation, the bleed air first flows through the distribution ducts in the wing leading edge and then to the leading-edge devices and is exhausted overboard. Figure 6-8 shows a typical wing anti-ice system.

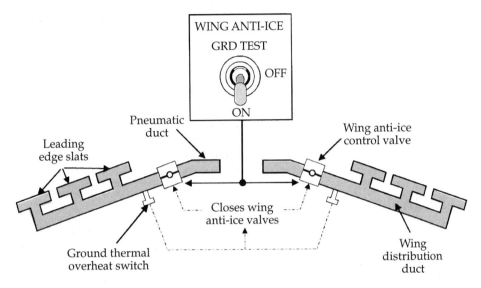

Fig. 6-8. A typical wing anti-icing system includes electrically operated valves to actuate the anti-icing with thermal overheat switches to protect the system.

Engine anti-ice is controlled from the cockpit by the engine anti-ice switches. Bleed air is routed to the inlet guide vanes, the nosecone and P_{t2} temperature probe, and the inlet cowling lip. Engine anti-ice is normally activated when the air temperature drops below a predetermined value, which is usually *above* the freezing point of water, and visible moisture is present. Engine anti-ice must be activated above freezing due to the temperature drop through the engine inlet.

Cockpit-window anti-icing

Cockpit window anti-icing can be accomplished with engine bleed air or by the use of electrical heating elements. Temperature sensors and temperature controllers maintain the window temperature in the range that the manufacturer has determined ensures maximum strength to the windows against the impact of birds in flight. The window heat also provides for window defogging.

Rain repellent and rain removal

Rain repellent is used in conjunction with the windshield wipers to improve visibility during heavy rain. When actuated, a predetermined amount of repellent from a disposable container is sprayed on the captain's and first officer's windows. The solution is spread over the windows by the rain flowing over the windows and by the action of the wiper blades. This provides the windows with a water-repellent coating.

The windshield wipers are provided to maintain a clear area on the pilot's windows during takeoffs, approaches, and landings in rain or snow. Each wiper is typically operated by a separate system to ensure that at least one window will be clear in the event of a system failure. The wiper blades are stowed in a park position when not in use.

Electrical anti-icing

The pitot probes, static ports, total-air temperature probe, and alpha vane, or angle-of-attack sensor, are electrically anti-iced. If the ground shift or ground sensor switch does not interrupt the power to the pitot-static, temperature, and alpha sensors, there is typically a time limitation placed on this heat on the ground. These probes, ports, and vanes are anti-iced during all flight conditions in jets, not just when icing is anticipated.

Endnotes

1. One of the early skid warning systems was found on the original Cessna Citation 500s. The release signal on this system was a "foot thumper," which was a small dc electric motor attached to a small peg that protruded through the pilot's rudder pedals. When an approaching skid was sensed, the motor was activated to rapidly tap the peg against the pilot's foot. The surprise of it all usually caused the pilot to pull his or her feet back, which effectively controlled the skid.

2. Aircraft designed by Ted Smith are a notable exception. The Aerostar and Aero Commander aircraft both have unconventional means of nosewheel steering. Both are hydraulic, but the Aerostar uses a rocker switch on the throttle pedestal while the Commanders use the toe brakes. Just slightly tapping the toe brakes activates the steering. Tapping the toe brake just slightly more activates braking. It takes some getting used to.

3. A French-made single-engine aircraft, the Rallye Minerva, had leading-edge slats automatically deployed in response to changing airflow and pressure over the wing. It was fascinating to change the angle of attack of the wing and watch the wing change shape.

7

Flight directors, autopilots, navigation systems, and performance-computer systems

FLIGHT DIRECTORS

The flight director is one of the better inventions of modern times. Students are usually resistant to use of the flight director and I'm not sure if this is a bit of bravado—"I don't need those crutches"—or just a misunderstanding of what the system really does. A good flight director is an invaluable aid to have when learning a new aircraft because it will display the proper pitch and roll commands to put the airplane where you have told the system you want to be. A truly professional pilot will use all the aids available in the cockpit to help get the job done efficiently and safely, and the flight director is certainly a tremendous aid.

Flight director indicator

The flight director indicator displays pitch and roll commands to capture and track a preselected heading or radio signal or to maintain a constant altitude, desired pitch attitude, rate of descent, or preselected indicated airspeed. Pitch and roll attitude are indicated by a horizon bar and an attitude tape referenced against a fixed airplane signal; the display can be thought of as a view of the aircraft as seen from the tail looking forward. Pitch and roll commands are indicated by the inverted V-bar command bars referenced against the fixed airplane symbol. The airplane, represented by a delta-shaped symbol, need only be flown in close formation with the inverted V command bars to satisfy the command signal.

The flight director indicator, also called the attitude director indicator or ADI, presents what at first can seem to be an overwhelming amount of information for the pilot. Most ADIs on jets are quite large. Because one degree of pitch change can make a large difference in jets, the ADIs are larger than those found in other aircraft in order to make one degree discernible.[1] Figure 7-1 shows a representative ADI. This ADI discussion will regularly refer to Fig. 7-1. This is meant to be a

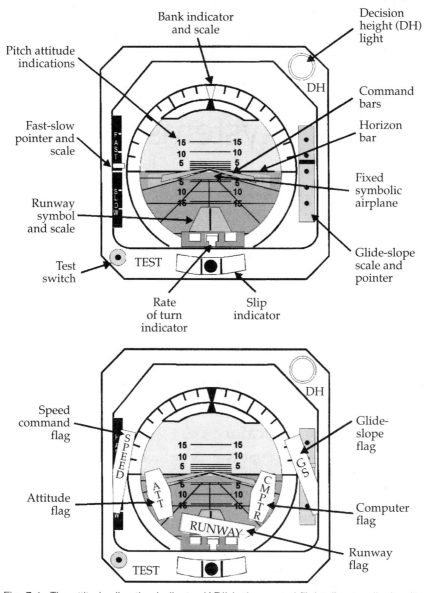

Fig. 7-1. The attitude direction indicator (ADI) is the central flight-director display. Its various indicators and failure flags are discussed in the text.

generic discussion of flight directors, not an explanation of any one model. General operating principles are the same across platforms.

Attitude assembly The attitude assembly includes the attitude tape, the horizon bar, and the bank pointer and scale. The attitude tape moves vertically in response to signals from the vertical gyro to display pitch attitude information from 90° pitchup to 90° pitchdown. The pitch attitude is scaled in increments of 1° with wider marks at 5° intervals up to 15°. The horizon bar moves in pitch and roll to display the position of the airplane in relationship to the horizon. The bank pointer moves against a fixed scale in response to signals from the vertical gyro. The fixed scale indicates angle of bank in increments of 10° for up to 360° bank.

Fixed symbolic airplane This fixed delta-shaped symbol displays aircraft attitude when related to the horizon bar.

Command bars The command bars are biased out of view with the flight director mode selector in the OFF position or when any of the warning flags appear. These inverted V-shaped command bars display pitch and/or roll commands as selected by the mode selector to track selected headings and to capture and track a radio course, glide slope, or preset pitch command. The pilot need only align the airplane signal with the command bars and fly close formation with them.

Decision-height light (amber) This light illuminates when the radio altimeter pointer reaches and descends below the selected MDA cursor setting. An aural tone will also be heard.

Rate of turn indicator The rate of turn indicator displays the rate of turn in increments of 3° for one needle-width deflection.

Slip indicator The ball monitors slip or skid for coordinated flight.

Glide-slope scale The glide-slope scale is a fixed scale with markings indicating glide-slope deviations of 0.35° for one dot and 0.70° for two dots. This scale is covered by the glide-slope flag when that flag is in view.

Glide-slope pointer The glide-slope pointer displays the vertical position of the airplane in relation to the glide slope. This pointer is out of view until the localizer is captured. The glide-slope pointer is in parallel with the glide-slope bar on the HSI.

Runway symbol and scale The runway symbol is in view when a reliable localizer signal is being received. It represents the position of the localizer beam relative to the runway. Outside marks on the scale represent 1¼° deviation from the localizer centerline and correspond to one dot deviation on the HSI course-deviation scale. At 200 feet radio altitude during an ILS descent, the runway symbol starts moving upward toward the airplane symbol and is aligned with the bottom of the symbolic airplane when the radio altimeter reads zero.

Fast-slow pointer and scale The fast-slow pointer moves in response to speed-computer signals when auto throttles are installed. In some flight director installations, this pointer is coupled to the angle-of-attack system. The scale will vary with the aircraft, but the center position of the pointer when coupled to the angle of attack represents V_{REF} or 1.3 V_{S1} for the flap configuration selected.

Test switch Depressing the test switch causes the attitude display to indicate a climbing right turn; 10° pitchup and 20° right roll are shown. The ATT and CMPTR flags should also appear, and the command bars should be driven from view.

Glide-slope flag The glide-slope flag monitors the glide-slope signal and glide-slope receiver when a localizer frequency is tuned. The flag will be in view if the CMPTR flag is in view during an approach, if the glide-slope signal falls below a predetermined level or if the glide-slope receiver fails. This flag is in parallel with the glide-slope flag on the HSI.

Attitude flag The attitude flag will be in view if power is lost to the vertical gyro or if the reference signal is unreliable. If this flag is in view, the CMPTR flag will also be in view.

Runway flag The runway flag is in view when a localizer frequency is tuned and the signal from the navigation receiver is unreliable or fails. If this flag comes in view after localizer capture, it causes the command bars to retract. This flag is biased out of view with a VOR frequency tuned.

Computer flag This flag monitors the attitude flag and the command bars display. It is in view if the attitude flag is in view or if the steering computer information is not valid or if the computer has failed. The command bars will retract when this flag is in view.

Horizontal situation indicator (HSI)

The horizontal situation indicator (HSI) will probably be more familiar to most pilots than the ADI since many general aviation aircraft used for instrument training are equipped with an HSI. The HSI depicts the position of the aircraft in relation to a selected navigation signal. The compass card on the HSI displays the magnetic heading as received from the remote-compass (flux gate) sending unit.

Figure 7-2 shows a representative HSI. Refer to this illustration during the following discussion about a generic HSI. Again, this is not meant to be faithful to any make or model, but is rather a generic discussion of HSIs.

Fixed-airplane symbol The symbolic airplane is fixed in the center of the instrument face and displays the position of the aircraft in relation to the movable parts of the indicator.

Lubber line The lubber line is fixed at the 12-o'clock position on the instrument. The aircraft heading is read on the azimuth card under the lubber line.

Azimuth card The azimuth card, also called a compass card, is slaved to the flux gate compass and displays aircraft heading under the lubber line.

Course pointer The course pointer is rotated in respect to the azimuth card by the course selector. The course corresponds to the digital readout in the course counter.

Course selector Rotating the course selector causes the course pointer on the HSI to move with relation to the azimuth card. The same course reference is provided to the flight director computer, and commands are then displayed on the ADI to fly and maintain the selected course when the VOR/LOC function is selected on the flight-director mode selector panel. A digital readout of the selected course is displayed in the course window on the HSI.

Course indicator A digital readout of the VOR radial or localizer course is selected with the course selector. This readout corresponds to the position of the course pointer.

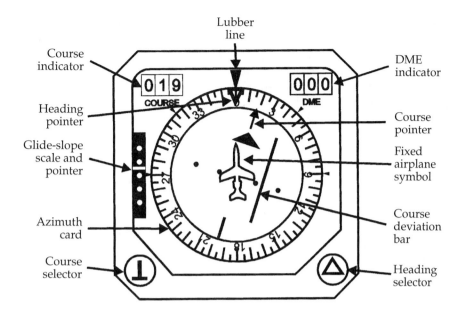

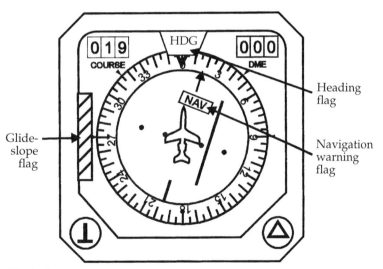

Fig. 7-2. The horizontal situation indicator (HSI) gives course, course deviation, and heading information to the pilot.

Course-deviation bar The course-deviation bar is the movable center portion of the course pointer and represents a segment of the selected VOR radial or localizer path. Rotation and lateral movement with respect to the fixed aircraft reference shows direction and amount of airplane deviation from the selected course.

Heading pointer The heading pointer moves with respect to the azimuth card when the heading selector is rotated. Once set, the heading pointer rotates with the azimuth card. The heading pointer will be under the lubber line if the selected heading is being held.

Heading indicator The heading indicator is a digital readout of the heading selected with the heading selector. The heading indicator corresponds to the position of the heading pointer.

Heading selector Rotating the heading selector moves the heading pointer on the HSI with relation to the azimuth card and provides the same heading reference signal to the flight-director computer. Commands are then displayed on the ADI to fly and maintain the selected heading when the HDG function is selected on the flight director mode selector panel. A digital readout of the selected heading is displayed in the heading window on the HSI.

Glide-slope scale The glide-slope scale is a fixed scale with markings indicating 0.35° deviation from the glide slope for one dot and 0.70° deviation for 2 dots. The scale is covered by the glide-slope flag when it is visible.

Glide-slope bar The glide-slope bar displays the vertical position of the aircraft in relation to the glide slope. It is in parallel with the glide-slope pointer on the ADI.

TO/FROM flag The flag indicates direction to VOR station along radial selected with course pointer.

Heading flag The heading flag will be in view if ac power is lost to the instrument, if heading information from the remote compass is lost, or, in the case of the heading card sticking, when the card and the signal from the remote compass disagree by more than 4°.

Navigation warning flag This flag will be in view if the VOR or localizer signal is weak or fails or if the NAV unit malfunctions.

Glide-slope flag The glide-slope flag will not appear unless a localizer frequency is selected. The flag will be in view when a localizer frequency is selected and the glide-slope receiver power fails, the glide-slope receiver fails, or the glide-slope signal is not at a usable reception level. The computer flag on the ADI will be visible if the approach mode is selected on the flight director mode selector and the glide-slope flag is visible.

Flight director mode selector/mode annunciator

The flight director mode selector panel allows the pilot to select the reference signals that the steering computer will use to compute steering commands and transmit them to the command bars. The flight director mode annunciator displays the mode that the flight director is actually following. In other words, the selector panel tells the flight director what the pilot wants, and the mode annunciator tells the pilot what the flight director is actually doing. The flight director mode annunciator in this case is combined with the autopilot mode annunciator. An illustration of a representative mode selector panel and mode annunciator is shown in Fig. 7-3.

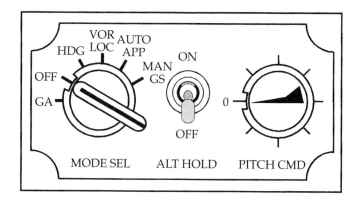

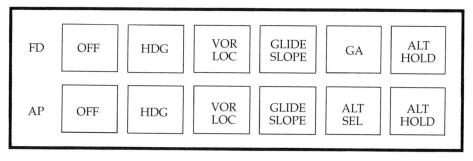

Fig. 7-3. The flight director mode-selector panel allows the pilot to program the flight director for the desired modes. The flight-director annunciator indicates the commands that the flight director is currently following.

Altitude-hold switch This switch, in the ON position, references altitude information from the flight-data computer for pitch commands. The switch is spring-loaded to the OFF position and held in the ON position by a solenoid. Altitude hold is effective in HDG, VOR/LOC, and AUTO APP prior to glide-slope capture. After glide-slope capture, the switch defaults to the OFF position. Altitude hold is not effective in the go-around mode.

Pitch command The pitch-command knob is used to set fixed angles of climb or descent in the range of 10° pitchdown to 15° pitchup. This command is not effective with the altitude-hold switch ON, after glide-slope capture, or in go-around mode.

Mode selector The mode-selector knob is positioned to determine which reference signals the steering computer uses to compute and transmit steering commands to the command bars. A solenoid holds the selector in position. The modes selectable on the mode selector are:

- OFF: Command bars are driven out of view. The ADI in this mode is an attitude display only. The red OFF light is illuminated on the annunciator display.

- HDG: Computes and commands in roll channel to fly and maintain airplane on the heading selected by the heading selector. May be used with altitude hold or pitch command. The green HDG light is illuminated on the mode annunciator.

- VOR/LOC: Computes and transmits commands to maintain airplane on selected VOR radial or localizer course with crab angle as required using selected radio and course deviation information as references. Prior to VOR/LOC beam capture, the computer will transmit commands to maintain the heading as selected by the heading selector. The VOR/LOC light will illuminate amber to indicate it is in the armed mode. At 5° (one dot in VOR or 2½°, 2 dots, in LOC) from beam center, the computer goes to beam-capture mode. The VOR/LOC will illuminate green to indicate it is in the capture mode, and the command bars will command the proper turn to the selected course.

- AUTO APP: Follows computed commands to maintain airplane on selected localizer course with crab angle as required using selected course and radio deviation information as references. Altitude hold or pitch command may be used with this mode. Prior to localizer-beam capture, the VOR/LOC light will be illuminated in the amber, armed mode, and the flight director will follow heading commands as selected in the heading selector. At 2 dots –2½° from LOC beam center, the computer goes to beam capture and the VOR/LOC light will illuminate green indicating capture mode. The glide-slope light will illuminate amber to indicate the armed mode. Approaching the glide-slope beam center (approximately ½ dot), glide-slope capture occurs. The glide-slope light, in addition to the VOR/LOC light, will be illuminated green to indicate capture. Altitude hold spring loads to OFF, if on, and a command is given to pitch the airplane down. Computed commands command a pitch attitude to follow the glide-slope center.

- MAN GS: This is a detented position. The VOR/LOC and glide-slope lights will be illuminated green. A computed command will be given to immediately pitch to the glide-slope beam. A fixed angle of intercept to the localizer will be commanded. After beam capture, this mode will be the same as AUTO APP.

- GA: The go-around mode is initiated by depressing the switches on the aircraft throttles if the mode selector is in AUTO APP or MAN GS. The go-around light will illuminate green and the mode selector will spring load to the GA position. The computer will transmit commands for the proper pitchup attitude and bank wings level. This mode may be initiated from any flight director mode by placing the mode selector in the GA position. To cancel this mode, select any other mode with the selector.

Using the flight director

The flight director can be used from takeoff to crossing the runway threshold. When cruising at altitude, the flight director commands the autopilot. The autopilot annunciator, combined with the flight-director annunciator, indicates the commands being followed.

Takeoff

Prior to takeoff, GA is selected on the flight director mode selector. This selection will drive the command bars to a wings-level pitchup that will approximate the pitchup required to maintain V_2 in the event of an engine failure after V_1. As soon as the pilot has determined that no engine failure has occurred, the pitch command with the appropriate pitchup for the climb desired and HDG can be selected on the mode selector. Figure 7-4 shows the flight director and HSI set up for a takeoff. The go-around mode has been selected on the flight director. A course of 034 has been selected on the HSI. Notice that the command bars have commanded approximately an 8° pitchup with wings level. At V_R, the symbolic airplane will be rotated to the command bars.

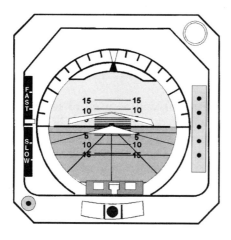

Fig. 7-4.
The flight director is programmed for a go-around and is commanding a noseup pitch attitude.

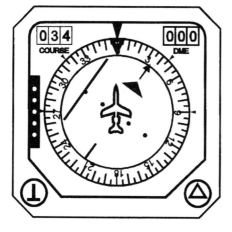

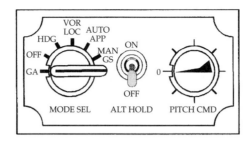

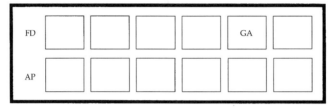

Fig. 7-4. Continued.

Climb/en route

During climb and en route, the flight director will be used to select and command the proper pitch for the rate or airspeed in the climb. The heading and navigation modes will be used for course selection. Figure 7-5 illustrates a climbing left turn to a heading of 300° to intercept the 330° bearing to the selected VOR.[2]

When the assigned altitude is reached, the aircraft will be leveled and altitude hold selected. If the autopilot is on and the altitude has been preselected, the autopilot will provide a parabolically smooth level off, computing its lead from the rate of climb, and will switch from altitude select (ALT SEL) to altitude hold. With the autopilot engaged, actuating electric elevator trim or pressing the autopilot/trim disconnect button on the yoke will disengage the autopilot and activate an aural tone and warning light.[3]

Approach

Flight directors are a tremendous aid on an instrument approach. Of course they can be coupled to the autopilot, but for training, hand flying with flight-director commands will give the student pilot excellent guidance and training. Figure 7-6 shows a flight director programmed for an ILS approach. The localizer has been captured, but the aircraft has drifted slightly right of course. Notice the command bars commanding a turn to reintercept the localizer centerline. The runway symbol is also to the left of the symbolic airplane. The glide slope is alive but is still two dots high. When the center of the glide-slope beam is captured, the flight director command bars will command the appropriate downward pitch to maintain centerline.

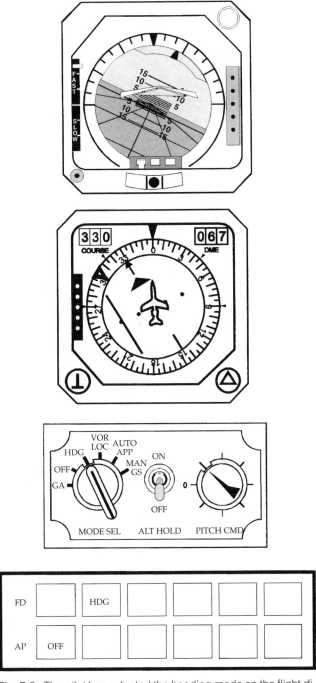

Fig. 7-5. The pilot has selected the heading mode on the flight director and has selected a heading of 300 on the HSI. The ADI
shows the aircraft in a climbing left turn in response to the flight-
director commands.

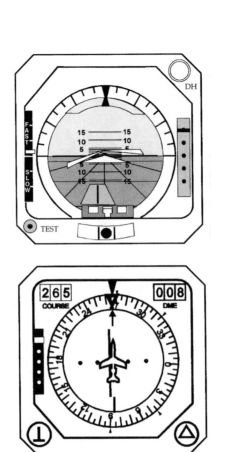

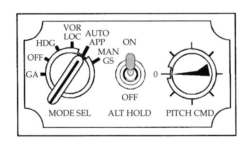

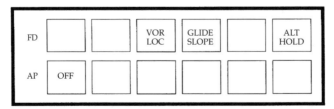

Fig. 7-6. The flight director is programmed in the approach mode. The command bars are commanding a left turn to return to the centerline of the localizer. The HSI shows the localizer needle to be slightly to the left.

AUTOPILOTS

An automatic flight control system typically consists of two subsystems: the stability augmentation subsystem and the autopilot subsystem. These two subsystems function independently or as parts of the total system.[4]

Stability augmentation subsystem

The stability augmentation subsystem is itself comprised of two additional subsystems. The Mach-trim compensator mode is established when the MACH TRIM COMP switch is in the normal position and power is applied to the system. Pitot and static pressures, supplied by the air-data computer, control the Mach trim actuator output in relation to Mach number to compensate for the nosedown pitching moments generated during operation at high Mach numbers.

The yaw-damper system is activated any time the autopilot is engaged, regardless of the position of the yaw damp control switch. The yaw damper may be engaged independently of the autopilot by selecting the ON position of the yaw damper control switch. The yaw-damper computer receives signals from the yaw-rate gyro and, if the autopilot is engaged, from the roll-axis computer. The yaw-damper computer sends signals to the rudder to effectively dampen any yawing tendency in either manual or automatic flight. The rudder movements generated by yaw-damper action are not transmitted to the rudder pedals.

Automatic-pilot subsystem

The automatic pilot controls the airplane vertical speed, altitude, and heading to reduce the workload on the flightcrew and to provide improved flight comfort and stability. The autopilot consists of a roll channel and a pitch channel. These, together with the yaw damper, control the aircraft around the pitch, roll, and yaw axes.

The automatic pilot stabilizes the aircraft in pitch and roll axes and steers it to headings set in manually by the pilot or transmitted to it by the compass system or from navigational radio signals. Pitch attitude may be controlled to maintain a precise barometric altitude, a fixed-pitch angle of climb or descent, or a fixed vertical speed.

The autopilot receives signals for the roll axis from several sources. Heading displacement signals are received from the directional gyro, roll displacement from the vertical gyro, navigational information from the VOR/LOC receiver, and heading select and course error information from the compass and flight director system. The pitch axis receives basic pitch displacement information from the vertical gyro, glide-slope information from the glide-slope receiver, and altitude information from the air-data computer.

Automatic pilot controller

Vertical-speed wheel Figure 7-7 illustrates a typical autopilot control panel. The VERT SPEED wheel is the primary pitch control for the autopilot and will

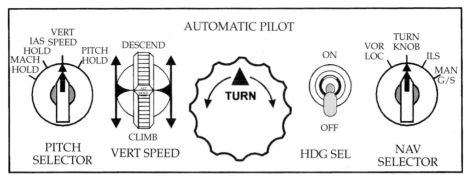

Fig. 7-7. Autopilot modes are selected through the autopilot control panel.

maintain the aircraft vertical speed that is existing when the autopilot is engaged. With the wheel in the ALT HOLD detent, the autopilot will hold the airplane at the barometric altitude that was existing when the autopilot was engaged. When the autopilot is not operating in the altitude preselect mode, movement of the wheel from the detent when under control of any other vertical mode will return the autopilot to the vertical-speed mode. When altitude preselect is on, the vertical-speed wheel is under control of the altitude preselect controller and will return to the detent when the preselected altitude is intercepted.

Pitch-selector knob The pitch-selector knob permits the pilot to select any of the specific pitch axis commands desired. Altitude preselect operation can be accomplished in MACH HOLD, IAS HOLD, VERT SPEED, and PITCH HOLD modes. When in the VERT SPEED mode, the rate of climb or descent is established with the vertical-speed wheel.

The MACH HOLD mode uses Mach airspeed information from the air-data computer to automatically maintain the indicated Mach number that existed when MACH HOLD was selected on the PITCH HOLD selector. IAS HOLD mode works in the same manner as MACH HOLD, but will maintain an indicated airspeed that was existing when the IAS HOLD mode was selected on the PITCH HOLD selector.

Selection of the PITCH HOLD function will command the airplane to maintain the pitch attitude that was existing when the PITCH HOLD mode was selected. This mode references the vertical gyro to obtain information for operation.

Turn knob The TURN knob establishes basic directional control over the airplane. Rotation of the knob will command a coordinated turn in the direction of rotation. The bank angle maintained will be proportional to the degree of knob rotation. The TURN knob takes priority over all other directional mode commands. When the TURN knob is returned to the center detent position, the autopilot will maintain the heading that existed when the bank angle has decreased to approximately 5 degrees on the roll out from the turn.

Heading select The HDG SEL switch establishes the heading mode and the autopilot will steer the airplane along the heading selected on the HSI. The airplane will be turned through the smallest angle between the present heading and

the selected heading and will maintain that heading until a new heading is selected or the heading select mode is disengaged.[5] The HDG SEL switch adjacent to
the TURN knob engages the heading-select mode when actuated and when the
TURN knob is in the center detent. Heading select may be used during the armed
phase of radio-beam navigation, but will be automatically deselected when the radio beam is intercepted.

Navigation selector The NAV SELECTOR switch provides for selection of
radio navigation signals or TURN knob input for the autopilot. When the selector
switch is placed in the VOR/LOC position, the autopilot will automatically capture and track VOR or localizer signals selected on the navigational radios. VOR
radial heading, corrected for wind, is provided to the system during VOR station
passage. The autopilot will not commence beam interception or tracking until the
TURN knob is in the center detent.

Altitude preselect

The altitude preselect mode provides the capability to preselect a desired altitude
and to perform either a manual or automatic level-off maneuver when the preselected altitude is reached. The altitude at which the level-off begins is determined
by vertical speed and the preselected altitude. During both manual and automatic
modes, the airplane's vertical rate is determined by the pilot's input.

When the altitude preselect is in the automatic mode, the autopilot will automatically perform a smooth level-off and capture the preselected altitude to within
approximately 100 feet. In the manual mode, the preselected altitude light located
on each pilot's instrument panel will illuminate when the airplane reaches the
level-off altitude as determined by the autopilot. This light informs the pilot to
perform the level-off maneuver.

When the altitude preselect mode is used during an ILS approach, the glide
slope will take priority over altitude preselect. When the glide slope is intercepted,
the altitude preselect switch will return to the OFF position (Fig. 7-8).

NAVIGATION SYSTEMS

Navigation systems can range from simple VOR/DME systems through the latest
GPS. Loran and RNAV receivers are common in general aviation aircraft, and most
pilots who have obtained an instrument rating have at least had ground-school fa-
miliarization with these systems if not used in flight. GPS, the global positioning
system, is also becoming very popular in the training fleet. ONS, the Omega navigation system, is found almost exclusively in jet aircraft.

Omega navigation system

The Omega navigation system provides continuous position and waypoint navigation displays computed by controlled comparison of very low frequency (VLF)
radio signals. These signals are transmitted by eight transmitting stations providing worldwide coverage. The eight stations share only three frequencies, but each
station transmits signal bursts with precise signal and time intervals. This enables

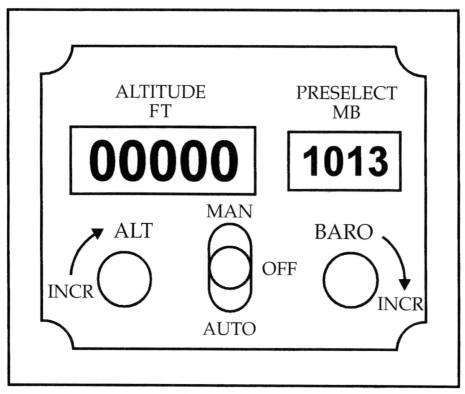

Fig. 7-8. The altitude preselect panel allows the crew to preselect altitudes for autopilot capture and hold.

the ONS computer to recognize each station's transmission. It is then possible to measure signal phase synchronized with the transmission start time.

The airborne components of the ONS are the antenna coupler unit (ACU), the processor receiver unit (PRU), and the control display unit (CDU). The installation might consist of a single unit or dual units. ONS computes aircraft position each 10 seconds when two or more Omega stations are receivable. ONS differs from loran and INS in that it does not measure elapsed time between master and slaved station signals, nor does it sense spatial acceleration; therefore, no gyro platform nor accelerometers are needed.

It is necessary to enter accurate time and date information into the ONS during the initialization process to enable the ONS to compute the ionosphere effect for each receivable Omega station. The propagation of VLF signals is altered by the ionosphere. Because the height of the ionosphere changes with conditions of daylight or darkness, climate, temperature, and season, it is necessary for the ONS to know the time and date information in order to make the proper calculations and corrections. In addition to time and date information, the starting position must be accurately entered into the ONS during initialization. The system may be initialized on the ground or in flight any time accurate position information is known.

The ONS computer will reject any Omega signals that are not suitable for navigation. Only two stations are required to determine position. Should two good stations not be available at any time, the ONS will revert to a dead reckoning navigation mode and will illuminate the DR annunciator light to alert the crew of this fact. When two good signals are again available, the ONS will automatically revert to Omega navigation and the DR annunciator will be extinguished. Normal accuracy of the Omega mode is generally within 1–2 miles for the entire flight.

Ground station malfunction or maintenance tests might produce faulty transmissions that will not be detectable by the ONS computer. For this reason, it is important that the crew maintain up-to-date station status notices. Stations that are off for maintenance can then be manually deselected by the crew.

A typical ONS system control panel and display is shown in Fig. 7-9. The functions, controls, and displays found on this receiver are:

- Left and right data displays. These LED displays display data as selected by the data-display selector.
- XTK key. The XTK—cross-track—key activates the process for manually selecting a track parallel to the desired track.
- DIM. The DIM controls the brilliance of the data displays.
- WPT. The waypoint selector selects the waypoint number for which the coordinates are displayed in the data displays or for which coordinates are to be entered.
- TO WAYPOINT display. The to-waypoint display contains the number of the waypoint for the navigation leg in use or for ranging calculations.
- TK CHG. The track-change key starts the process for manually changing the active navigation leg.
- MODE SELECTOR. The mode selector knob allows selection of the following system operating modes:
  ~ OFF: The system is not powered. The knob must be pulled and rotated to select other modes.
  ~ A: Automatic. Navigation legs are automatically sequenced for use, but can be manually changed.
  ~ M: Manual. Navigation legs must be manually changed for use.
  ~ R: Remote. Provides ranging calculations between selected waypoints.
- DISPLAY selector. The display selector selects data for presentation in the data displays and allows manual inputs for the dead-reckoning mode. The display selector has the following positions:
  ~ WPT: The latitude to one-tenth minute is shown in the left display and the longitude to one-tenth minute is shown in the right display. The waypoint number selected with the waypoint-number selector appears in the TO waypoint display.
  ~ DIS-TIME: The distance from the airplane to the TO waypoint is shown in the left display, and the estimated time en route to one-tenth minute is shown in the right display based upon the present ground speed. If

GMT is pressed, the right display changes to the estimated time of arrival referenced to GMT.

~ WIND: The true wind direction is shown in the left display; the wind speed is shown in the right display. The wind speed will indicate zero if the TAS is lost or less than 180 knots.

~ DTK/STS: The magnetic desired-track angle for the active navigation leg is shown in the left display. System status and action or malfunction codes are shown in the right display.

~ MH/TAS: Selection of this position shows magnetic heading to one-tenth degree in the left display and true airspeed in the right display.

~ STA: The Omega station status is displayed successively from left through right displays by station number.

~ AUX: This position provides for manual start of relaning, manual ground speed entry, display of memory, display of station frequencies in use, and display of VLF station status.

~ GMT/DAT: The left display shows Greenwich Mean Time to the nearest minute and the right display shows the date in order of month, day, and year.

~ TK/GS: The present magnetic track to one-tenth degree is shown in the left display and the ground speed is shown in the right display.

~ HDG/DA: The present magnetic heading to one-tenth degree is shown in the left display and the drift angle (left or right) is shown to one-tenth degree in the right display.

~ XTK/TKE: Crosstrack distance (L or R) to one-tenth mile is shown in the left display and track-angle error (L or R) to one-tenth degree is shown in the right display.

~ POS: The present aircraft position latitude to the nearest tenth of a minute is shown in the left display and the current longitude to the nearest tenth of a minute is shown in the right display.

- ENT key. The enter key transfers data entered from the displays and keyboard into the computer. This key illuminates to indicate that the computer is ready to accept data.

- HLD key. The hold key stops the present-position displays from changing, permits position update, selects the action/malfunction code displays, and permits editing of waypoints and entry to AUX functions.

- DTK key. The desired-track key permits manual selection of a rhumb-line track from the present position.

- MODE and WARNING annunciators. The mode and warning annunciators alert the crew to the operating modes in use and warnings pertinent to the ONS system. The various annunciations are:
  ~ ALR: Alert. This light will illuminate steady for 2 minutes before arriving at the next waypoint. If in M mode, the light will begin flashing one-half minute before arrival at the waypoint. The light will continue flashing until the track is manually changed or until the A mode is selected. This light will not illuminate if the ground speed is less than 180 knots.

- ~ DR: Dead reckoning. Illumination of this light indicates that the navigation computer is in the dead-reckoning mode.
- ~ VLF: Illumination of this light indicates that the computer is in the VLF navigation mode.
- ~ AMB: Ambiguity. The illumination of this light indicates that the system has detected a possible position error.
- ~ MAN: Manual. Illumination of this light indicates that true airspeed, magnetic heading, or crosstrack offset has been manually entered into the computer.
- ~ WRN: Warning. Steady illumination of this light indicates that the system is inoperative. When this light is flashing, it indicates that the system has lost a required input or has had a power interruption but is still operative.
- DATA KEYBOARD. The data keyboard provides 10 numerical keys for data entry.
- DAT key. The date key permits the entry of the date.
- CLR key. The clear key clears the displayed data if it has not yet been entered. This key also clears the system of entered erroneous data.

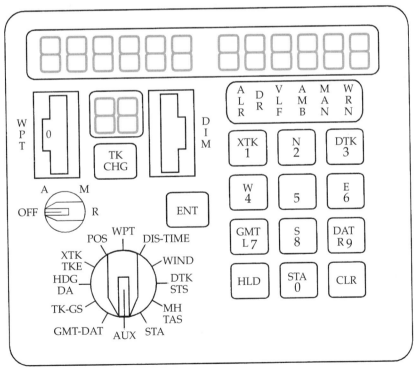

Fig. 7-9. The ONS-system control panel allows the crew to input data and annunciates operating modes and course information.

- STA key. The station key permits manual deselection of Omega stations and permits their reselection.
- GMT key. The Greenwich Mean Time key permits the entry of time.

Figure 7-10 shows two aircraft. The top aircraft is on track and the bottom aircraft is off track. The various items of navigation data are shown on the illustrations as follows:

- TK: Track angle. The angle, from magnetic north, of the current aircraft track. It corresponds to magnetic course.
- GS: Ground Speed. The ground speed being made good.
- HDG: Heading. The current aircraft heading in degrees magnetic used to make good the magnetic course.
- DA: Drift angle. The drift angle between the track angle and the aircraft heading. Corresponds to wind-correction angle.
- XTK: Crosstrack distance. The distance offset for the parallel track selected.
- TKE: Track angle error. The difference between the actual track angle and the desired track angle.
- POS: Position. The present position of the aircraft.
- WPT: Waypoint. The waypoint selected.
- DIS: Distance to go. The distance to go from the aircraft position to the waypoint selected.
- TIME: Time to go. The time to go from the aircraft position to the waypoint selected.
- WIND: Wind. The current wind direction and speed.
- DTK: Desired track angle. The track angle desired to proceed to the waypoint.

PERFORMANCE-DATA COMPUTER SYSTEMS

The performance-data computer system (PDCS) provides the crew with flight guidance data to assist them in achieving the most economical and efficient operation of the airplane. The system is controlled by the crew, and its components are the computer, a control-display unit, and a mode annunciator.

The primary function of the PDCS is to compute, from its various automated inputs, the targets, airspeeds, and power settings for takeoff, climb, cruise, descent, holding, and go-around. The PDCS computes and displays the optimum airspeeds and power settings on its display unit and also sets the power and airspeed bugs to the computed values. Additionally, the PDCS performs and displays, on request, various other calculations useful in flight planning and management, including range, trip fuel, reference speed, trip altitude, and wind.

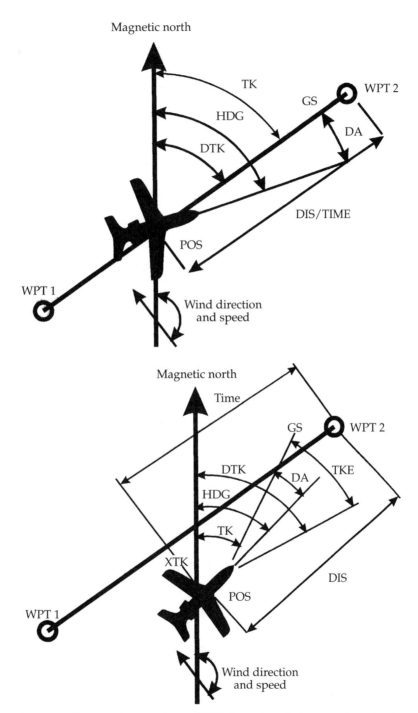

Fig. 7-10. The top diagram shows ONS information relative to a plane that is on track. The bottom diagram shows an aircraft that is off course. The various labels are discussed in the text.

Computer inputs

The PDCS receives the following inputs to its system for its computations:

- Temperature. Temperature is necessary for power, true airspeed, and reference-speed computations. The computer receives total air-temperature input from the air-data computers.
- Altitude and airspeed. Altitude and airspeed inputs are received from the air-data computer.
- Fuel weight. The total weight of fuel aboard the airplane is transmitted to the PDCS by a fuel summation unit that receives inputs from data transmitters in each of the airplane's fuel tanks.
- Bleed logic. The PDCS receives the switch position for all bleed-air switches and engine bleed-air configurations to adjust limiting power settings.
- Power setting. The existing power setting for each engine is transmitted to the PDCS for use in computing actual aircraft performance.
- Distance. The distance information used by the system is provided by the aircraft DME or the Omega receivers. This data is used in the computation of wind information and ground speed.

Computer outputs

The PDCS provides the crew with a variety of speed schedules for climb, cruise, and descent, enabling the crew to select the schedule that best suits their trip requirements. Climb schedules include a choice for economy climb (minimum cost), maximum rate of climb, or a speed entered by the crew.

During cruise, the crew is presented with economy, long-range cruise, or manual selections. Long-range cruise computes the best range speed, and economy computes the minimum-cost speed taking all parameters into account. The crew is also able to enter a desired speed. The computer will also provide a best speed for turbulence penetration.

Economy or manual descent choices are provided for crew selection for descent scheduling.

Economy climb and cruise schedules are computed to provide a minimum trip cost based upon a flight index entered into the computer. The flight index is a number between 0 and 200 with zero indicating that fuel economy is the exclusive criterion for selecting cruise speed. An index of 200 tells the computer that flight time is the overriding consideration and the speed will be computed to minimize the flying time at the expense of greater fuel consumption.

The PDCS has a complete built-in, self-test capability that allows a complete check of the computer and all of its inputs and outputs. If the PDCS fails, its screen will go blank. If the air-data computer fails, the PDCS screen indicates an error code.

Endnotes

1. It could also be that ADIs on jet aircraft are larger than those on other aircraft because by the time the pilot has the time and experience to make it to the right seat of a jet, it is usually necessary that instruments and printed matter be larger. I'm sure most jet pilots would not admit this (I sure wouldn't!), but I suspect it is true.

2. It is common to fly in the heading-select mode rather than VOR/LOC tracking. With some autopilots, it just seems smoother that way.

3. The autopilot disconnect warnings were required as a result of the Eastern Airlines Flight #401 that crashed into the Florida Everglades east of Miami after the pilot inadvertently disengaged the autopilot and put the aircraft into a slow descent. There was no warning that the autopilot had been disconnected.

4. Because this is a generic book, this autopilot discussion might not necessarily correspond to the flight director previously discussed. There are many different autopilot/flight director systems found on jet aircraft. The underlying operating principles of these varying systems are much the same with the main differences coming in the language used to name various modes.

5. Care should be taken when making turns of greater than 180 degrees to select less than 180 degrees for the initial turn and select the final heading when the amount of turn required to reach that heading is less than 180 degrees. For example, when turning left from 360 to 090, which is the long way around, select a heading of 200 initially, and when the airplane is turning through approximately 240, select 090 as the final heading. Selecting 090 initially will cause the autopilot to command a right turn to 090.

8

Performance calculations

PART 25 OF THE FEDERAL AVIATION REGULATIONS IS VERY SPECIFIC AS to the performance a transport category aircraft must demonstrate to be certified.[1] This performance must first be demonstrated and documented during certification flight testing, and then the information must be included in the official aircraft flight manual for use in determining runway requirements for takeoff and landing and obstacle clearance.[2]

FAR 91 requires the following of all pilots of turbine-powered aircraft:

91.605 Transport category civil airplane weight limitations.

. . .

(b) No person may operate a turbine engine powered transport category airplane certificated after September 30, 1958, contrary to the Airplane Flight Manual, or take off that airplane unless:

(1) The takeoff weight does not exceed the takeoff weight specified in the Airplane Flight Manual for the elevation of the airport and for the ambient temperature existing at the time of takeoff;

(2) Normal consumption of fuel and oil in flight to the airport of intended landing and to the alternate airports will leave a weight on arrival not in excess of the landing weight specified in the Airplane Flight Manual for the elevation of each of the airports involved and for the ambient temperatures expected at the time of landing;

(3) The takeoff weight does not exceed the weight shown in the Airplane Flight Manual to correspond with the minimum distances required for takeoff considering the elevation of the airport, the runway to be used, the effective runway gradient, and the ambient temperature and wind component existing at the time of takeoff; and

(4) Where the takeoff distance includes a clearway, the clearway distance is not greater than one-half of—

(I) The takeoff run, in the case of airplanes certificated after September 30, 1958, and before August 30, 1959; or

(ii) The runway length, in the case of airplanes certificated after August 29, 1959.

No person may take off a turbine engine powered transport category airplane certificated after August 29, 1959, unless, in addition to the requirements of paragraph (b) of this section—

(1) The accelerate-stop distance is no greater than the length of the runway plus the length of the stop way (if present); and

(2) The takeoff distance is no greater than the length of the runway plus the length of the clearway (if present); and

(3) The takeoff run is no greater than the length of the runway.

Definitions

Terms and abbreviations should be defined before we begin our discussion of FAR 25 performance.

Distances

Accelerate-stop distance is the total horizontal distance traveled from brake release to the point at which the airplane comes to a complete stop on a takeoff during which the pilot elects to stop at V_1. This distance is based on a smooth, dry, hard-surfaced runway.

Accelerate-go distance is the horizontal distance traversed from brake release to the point at which the airplane attains a height of 35 feet above the runway surface on a takeoff during which one engine fails at V_1 and the pilot elects to continue.

Takeoff field length is the distance presented in the takeoff distance charts and equal to the accelerate-stop distance or the accelerate-go distance, whichever is greater. If the accelerate-stop and the accelerate-go distances are equal, this distance is referred to as the *balanced field length*.

Speeds

IAS is indicated airspeed. The airspeed indicator reading assumes zero instrument error.

V_{MCA} is minimum control speed in the air, which is the minimum flight speed at which the airplane is controllable when one engine suddenly becomes inoperative and the remaining engine is operating at takeoff thrust.

V_{MCG} is minimum control speed on the ground, which is the minimum speed on the ground at which control can be maintained using aerodynamic controls alone, when one engine suddenly becomes inoperative, and the remaining engine is operating at takeoff thrust.

V_1 is critical-engine failure speed, which is the speed at which, due to engine failure or other causes, the pilot may elect to stop or continue the takeoff. If engine failure occurs at V_1, the distance to continue the takeoff to 35 feet will not exceed the usable takeoff distance. The distance to stop the airplane will not exceed the accelerate-stop distance. V_1 must not be less than V_{MCG} or greater than V_R.

V_R is rotation speed, which is the speed at which rotation is initiated during the takeoff roll to attain V_2 at or before 35 feet.

V_2 is takeoff-safety speed, which is the actual speed at 35 feet above the runway surface as demonstrated in flight during single-engine takeoff. V_2 must not be less than 1.1 times V_{MCA} nor less than 1.2 times the stalling speed in the takeoff configuration, nor less than V_R.

V_{S0} is stalling speed in the landing configuration.

V_{S1} is stalling speed in the specified gear/flap configuration.

V_{APP} is approach speed, which is 1.3 times stalling speed with the flaps in the approach configuration and the gear up.

V_{REF} is landing approach speed, which is 1.3 times V_{S0}.

Miscellaneous

ISA is international standard atmosphere.

Demonstrated crosswind is adequate control of the aircraft that was actually demonstrated during certification test flights with the crosswind component of tower-reported winds at the stated value. This value is not considered to be an aircraft limitation.

Wind is the wind velocities recorded as variables in the performance charts as the headwind or tailwind component of the actual wind as reported 20 feet above the runway surface.

Takeoff brake energy limit is the maximum takeoff weight at which an aborted takeoff can be initiated at V_1 and completed within the accelerate-stop distance using maximum-braking energy.

Landing brake energy limit is the maximum gross weight at which the airplane can be brought to a full stop within the computed landing distance using maximum braking effort.

Runway gradient is change in runway elevation per 100 feet of runway length. The values are given as positive for uphill gradients and negative for downhill gradients.

Climb gradient (expressed in percent) is the ratio of the change in height during a portion of the climb to the horizontal distance traveled in the same time interval.

Gross climb gradient is the climb gradient that the aircraft can actually achieve given ideal conditions, also expressed in percent.

Net climb gradient is the gross climb gradient reduced by 0.8 percent for two-engine airplanes, 0.9 percent for three-engine airplanes, and 1.0 percent for four-engine airplanes during the takeoff phase. During the en route climb phase, the gross climb gradients are reduced by 1.1 percent, 1.4 percent, and 1.6 percent, respectively. This conservatism is required by FAR 25 for terrain-clearance determination to account for variables encountered in service.

TAKEOFF CONSIDERATIONS

Students new to jet aircraft are sometimes overwhelmed by the number and seeming complexity of the calculations that must be made just to determine whether or not a takeoff can be legally (and, I'll add, safely) made. A thorough understanding of FAR 25 performance calculations, however, will enhance the understanding of all aircraft performance and will perhaps give the reader cause to more carefully consider takeoff calculations in all aircraft.

Flying a FAR-25-certified aircraft is like getting a guarantee or warranty of performance. This warranty is just like any other warranty, however, and it requires that certain conditions be fulfilled to keep the warranty in full force and effect. If the pilot abides by the limitations in the aircraft flight manual, the minimum performance stated in FAR 25 will be available.

Before we explore just what that minimum performance is, let us determine how takeoff distances are calculated for certification of FAR 25 airplanes. All the demonstrations and calculations for FAR 25 airplanes make some basic assumptions:

- Dry, hard-surfaced runway with no gradient
- New aircraft with new engines and new brakes
- Power set and stabilized before brake release

Refer to Fig. 8-1(A–C). Three distances are taken into account when charting the takeoff runway required for an FAR-25 aircraft. The first is the all-engines operating takeoff distance. The aircraft is positioned on the end of the runway, and the power is set with the brakes held. This is done to eliminate the variables that would be introduced if a rolling takeoff were made. When the engines have stabilized at takeoff power, the brakes are released, and the aircraft accelerates down the runway. V_1 passes with all engines continuing to operate, and the aircraft accelerates to V_R and is rotated to the takeoff attitude. The aircraft lifts off the runway and climbs to 35 feet above the runway elevation with gear down and flaps in the takeoff position. The total horizontal distance from brake release to the point where the aircraft is 35 feet in the air is measured. Because it is obvious that the all-engine takeoff distance is always going to be less than the distance obtained with an engine failure, this figure is multiplied by 1.15. This is shown in Fig. 8-1A.

The aircraft is returned to the end of the runway and fueled to the same gross weight as in Fig. 8-1A. The power is again set with the brakes held, and when the engines are stabilized, the brakes are released, and the aircraft accelerates down the runway. At V_1, the inevitable happens, and an engine suddenly fails. The pilot elects to continue the takeoff, the aircraft accelerates to V_R, and the pilot rotates the aircraft to the takeoff attitude. The aircraft lifts off and climbs to 35 feet above the runway elevation, again with gear down and the flaps in the takeoff position. The total horizontal distance from brake release to the point where the aircraft is 35 feet in the air is measured. This is recorded along with the distance obtained in A.

One more trial is made. Everything is done as in the preceding two examples except that this time when the engine fails, the pilot elects to abort the takeoff and stop the aircraft on the runway. This is accomplished by applying maximum braking effort. The total horizontal distance from brake release to the point where the aircraft is stopped on the runway is measured. This distance is recorded and compared to the distances obtained in A and B. Even with the multiplication factor of 1.15 times the flight-test distance, the distance computed in trial A will always be found to be shorter than either B or C. The takeoff runway required, then, will be

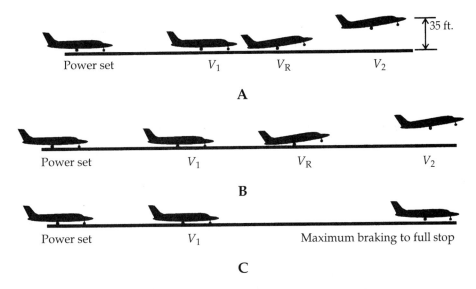

Fig. 8-1. The takeoff distance required is measured three times under FAR 25. The first test is done with both engines operating, and the distance measured is multiplied by 115 percent. The second test is done with an engine failure at V_1 and a continued takeoff. This distance is known as the accelerate-go distance. A final test is done and the takeoff is aborted at V_1. This distance is known as the accelerate-stop distance.

the longer of B (accelerate-go distance) or C (accelerate-stop distance). If these two distances are the same, the distance obtained is known as *balanced-field* length.

The myth of V_1

High-speed aborted takeoffs are one of the most dangerous maneuvers a jet pilot might be called upon to perform. A look at accident statistics or the national media will serve to confirm this. Many accidents are the result of a decision to abort a takeoff at a speed at or approaching V_1. Why should that be? A thoughtful look at takeoff performance will point out some areas where jet pilots are given a false sense of security.

First of all, an understanding of what V_1 *is* and what it *is not* should be gained. The definition of V_1 in FAR 25 is as follows:

(a) V_1 must be established in relation to V_{ef} as follows:

(1) V_{ef} is the calibrated airspeed at which the critical engine is assumed to fail. V_{ef} must be selected by the applicant, but may not be less than V_{MCG} determined under 25.149(e).

(2) V_1, in terms of calibrated airspeed, is the takeoff decision speed *selected by the applicant*; however, V_1 may not be less than V_{ef} plus the speed gained with the critical engine inoperative during the time interval between the instant at which the critical engine is failed, and the instant at which the pilot recognizes and reacts to the engine failure, as indicated by the pilot's application of the first retarding means during accelerate-stop tests.

The absolutely critical term here is the phrase "selected by the applicant." As can be readily seen from this demonstration, the only relation V_1 needs to have to actual aircraft performance is that it might not be less than V_{MCG}. Because runway-length requirements for takeoff are all based upon V_1—the higher V_1 is, the longer runway is required—you can bet that the applicant has the test pilot out there in the flight-test airplane repeating this test over and over until the very lowest number is obtained.

Second, remember that this important aircraft-performance number was obtained by a test pilot who does aborted takeoffs day after day. He or she is flying a brand-new airplane with brand-new brakes, and the engine failure—or other emergency—does not come as a surprise to him or her. That test pilot has rehearsed it more times than even the most thoroughly trained pilot could even imagine! The reaction time built into the equation is, consequently, somewhat false. Most pilots take approximately 4 seconds more to react than is allowed in the testing. At 120 knots, the aircraft will travel about 800 feet during those 4 seconds.

Third, and highly important, is that the manner in which the testing is conducted is not the manner in which jets are usually flown. Very rarely does a jet pilot sit on the end of the runway with the brakes held and run the engines to takeoff power and stabilize them prior to brake release. The jerky takeoff that results from this procedure is highly upsetting to passengers. This results in two factors that will cause the takeoff distances experienced in actual service to be far greater than the charted distances. The runway that is traversed while the power is being set to the computed takeoff-power figure must be added to the charted figures to get a more accurate takeoff distance.

Additionally, if power is set using an N_1 gauge, as is common for many fanjet engines, the maximum takeoff power will not be obtained. N_1 increases as the aircraft accelerates down the runway until about 60 KIAS is reached. This is also about the time that the power finally gets set on a rolling takeoff. The power that is finally set for takeoff can be as much as 2 percent less than that which would have been set in a static condition, which is what the takeoff charts (power charts and distance charts) assume. Because the top 30 percent of power in a fanjet engine is found in the top 10 percent of engine RPM, this can be a significant power loss that will serve to lengthen the takeoff distance.

The temperature figures used to calculate the takeoff distance are rarely those experienced on the runway. Most airports locate their thermometers over grassy areas. The runway temperature can be as high as 40°F warmer than that reported by the tower. This will have a significant detrimental effect on takeoff performance.

Any drag from control deflection introduced for crosswind correction will inhibit the acceleration to V_1. This is not taken into consideration when the charts are prepared. Only the headwind portion of any wind is allowed for. No correction factor is given for crosswind-correction drag.

All of these factors will serve to lengthen the runway required to accelerate to V_1. Now what about the runway required to stop in the case of an abort? Again, a major factor that causes a lengthening of this distance is the reaction time allowed during certification. This time is based upon the reactions of a test pilot who knows exactly what is going to happen and when it will happen. An actual pilot experiencing an engine failure or other emergency at or near V_1 will experience a period of shock and disbelief before reacting to the emergency.

Training, of course, will shorten this reaction time, but there again, we might have a somewhat false sense of security. When emergencies happen at V_1 in the simulator, most pilots are already springloaded to the ready position. We expect to have the sim instructor failing engines, blowing tires, deploying thrust reversers, and all other manner of unpleasantness. We're ready. We expect the worst. A two-engine nonemergency takeoff is the exception rather than the rule.

In the airplane on an actual takeoff roll, however, the situation is vastly different. No matter how much training a pilot has, in the real world, almost all takeoffs are successful. Few pilots have been faced with the necessity of doing a high-speed aborted takeoff. An expectation is formed that all takeoffs will be successful, and when that proves not to be the case, disbelief will lengthen the reaction time. Disbelief that a takeoff would not be successful is thought to be one factor in the Air Florida crash in Washington, D.C. It is surely a dangerous mindset for a pilot but one that is very difficult to eliminate.

When the shock and disbelief is over, the problem of stopping the airplane must be addressed. The certification tests are done with maximum braking effort. The test pilot is well aware what maximum braking effort is. Most pilots are not. Simulator studies done with pilots of all levels of experience showed that most of them, in the order of 80 percent, were not applying maximum braking effort when they thought they were. Maximum braking effort is a fairly extreme maneuver, and it is one that most pilots studiously avoid. Maximum braking effort pretty well assures that the brakes will not last through more than a few landings.

All these factors will tend to lengthen the accelerate-stop distance experienced in actual service. Of these, the less-than-maximum braking effort probably contributes the most to the added distance. Many pilots are unpleasantly surprised to find that the airplane takes almost twice the charted distance to stop after an aborted takeoff. The accelerate-go figures provided by the manufacturer are reasonably close to what can be achieved by the average pilot, but the accelerate-stop distances are woefully inadequate.

The first chart

The very first chart to be consulted when calculating takeoff numbers is a chart that is usually titled "Maximum Gross Weight for Climb Requirements" or some such similar wording. Recall that FAR 25 gives you some minimum performance warranties. It is up to you, however, to ensure that you do not void that warranty, and this chart is the first step in that direction. Refer to Fig. 8-2. This chart, which is representative of one from a popular corporate jet, lists the maximum takeoff weight allowed for ambient conditions. The figures obtained on this chart are limited by climb requirements. We'll discuss these climb requirements later, but first of all, let's see what this chart tells us.

First of all, the aircraft configuration upon which the chart is based is listed on the top of the chart: flaps at 20° and anti-ice off. The chart is entered on the left at the ambient temperature for takeoff. This example will use 70°F. Follow the 70°

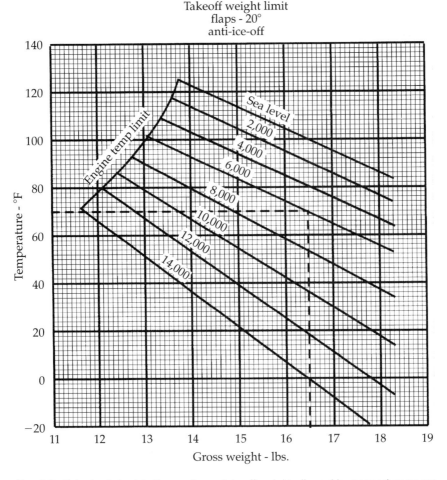

Takeoff weight limit
flaps - 20°
anti-ice-off

Fig. 8-2. This chart depicts the maximum takeoff weight allowed by second-segment climb requirements. The conditions of flap setting and anti-ice setting are crucial.

line to the right until it intersects the airport elevation curve, in this case, 6000 feet. Directly below the point of intersection is the maximum takeoff weight permitted for these conditions, 16,500 pounds. Limiting takeoff weight under these conditions will keep your warranty valid; the aircraft will achieve at least the minimum climb requirements of FAR 25.

This particular aircraft, however, has a structural takeoff gross weight limitation of 18,300 pounds, an 1800-pound reduction. That's several passengers, a lot of baggage, or reserve fuel. Is there any way to take off from this airport at this temperature? Luckily, there is another option. Notice the flap setting that this chart was predicated upon: 20°. Flaps ALWAYS hurt climb performance. True, they shorten the takeoff roll, but as soon as the wheels leave the pavement, the flaps become a liability.

This aircraft is certified for takeoff with a flap setting that results in less penalty, 8°, which in this aircraft is the minimum setting authorized for takeoff. Refer to Fig. 8-3. This chart lists the maximum takeoff weights allowed for ambient conditions when using only 8° flaps. Notice that with this flap setting, takeoff at maximum structural gross takeoff weight is allowed. Also notice that this chart has two parts, one labeled TAKEOFF CLIMB and the other labeled BRAKE ENERGY. Takeoff with a lower flap setting requires higher speeds, and because of this, brake energy required to stop the airplane in the event of a takeoff aborted at V_1 must be considered. At a 6000-foot airport elevation, this is not a problem, but notice that if the airport elevation is 8000, the takeoff climb weight permitted is still 18,300 but the brake energy weight permitted drops to 17,800. (This is not totally true because we have not taken headwind and runway gradient into account.)

We now know that the airplane will climb at some minimum rate once we get it off the runway after an engine failure at V_1, but now we need to determine if there is enough runway to accomplish the takeoff. Table 8-1 lists the takeoff runway requirements for 8° flaps. At 18,000 takeoff weight and 75°F ambient temperature, the total runway required is 10,750 feet. If you happen to be at Cortez, Colorado, you have a problem because the runway there measures 7205 feet. You can reduce the weight to 16,500 and takeoff with 20° flaps, or you can wait until winter and a temperature reduction!

Table 8-1. Takeoff distance—flaps 8°
6,000 feet MSL

Temperature °F	Gross Weight				
	18,300	18,000	17,500	17,000	16,500
−25	5,750	5,500	5,250	5,000	4,750
0	6,250	6,000	5,750	5,500	5,250
25	7,500	7,250	7,000	6,750	6,500
50	8,000	7,750	7,500	7,250	7,000
75	11,000	10,750	10,500	10,250	10,000

FAR 25 climb profile

The FAR 25 climb profile is the profile that the airplane had to demonstrate to be certified. It is certainly not the way normal takeoffs are flown, although that is hard to get across to some students and instructors alike. Examiners insist on seeing this profile flown on the V_1 cut takeoff to determine that the student understands what this profile consists of. It certainly is inappropriate on most other takeoffs, however.

As discussed earlier, the power is set with the brakes held for the FAR 25 certification profile. The brakes are then released, and the aircraft proceeds down the runway to V_1. Magically at V_1, there is an engine failure, and the pilot elects to continue the takeoff. The aircraft is accelerated to V_R, rotated, and lifted off the run-

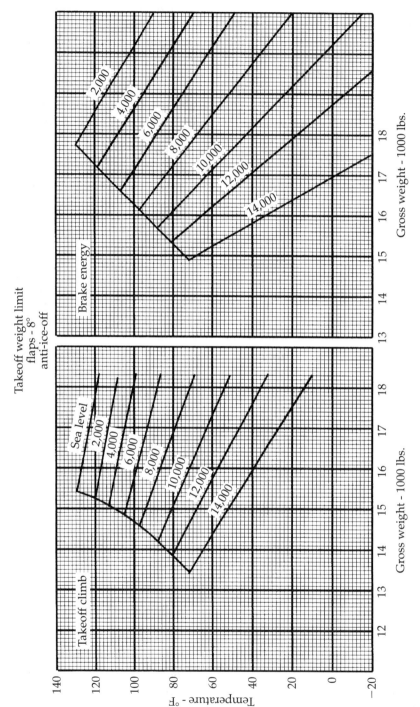

Takeoff weight limit
flaps - 8°
anti-ice-off

Fig. 8-3. This chart depicts the maximum takeoff weight allowed by second-segment climb requirements and by brake-energy limits in the event of an aborted takeoff.

way. With the gear down and the flaps in the takeoff position, the aircraft climbs to 35 feet above runway elevation. The aircraft speed at 35 feet should be V_2. This point is termed "Reference Zero" because this is where the takeoff distance ends and the FAR 25 takeoff climb begins. Again, the configuration at this point is:

- One engine windmilling
- One engine at takeoff thrust
- Flaps in the takeoff position
- Gear down
- Speed—V_2

Reference zero is the point where the first segment of the FAR 25 climb begins. The only thing that happens during the first segment is gear retraction. The aircraft must demonstrate a positive climb in the first segment to meet certification criteria.[3] This might seem strange at first, but remember that in some aircraft, landing-gear retraction can result in additional drag due to the opening of gear doors.

Appropriately enough, at the end of the first segment, the second segment begins. This climb segment limits the maximum permitted takeoff weight of the aircraft. The second segment begins at the altitude the aircraft has achieved after the first segment and ends when the aircraft reaches 400 feet AGL. The minimum gross climb gradient that the aircraft must demonstrate in this segment is 2.4 percent. That is 2.4 feet of climb for every 100 feet horizontal travel. That sure isn't much, but it is in the proper direction. (At a baseball diamond, a climb at that rate starting at home base will not clear the pitcher's mound.) The aircraft configuration in the second segment is:

- One engine windmilling
- One engine at takeoff thrust
- Flaps in the takeoff position
- Speed—V_2

Notice that the only difference in configuration between the beginning of the first and second segments is the position of the landing gear.

The second segment ends at 400 feet AGL. At this point, the third, or acceleration, segment begins. The aircraft is accelerated to a speed above V_2 that is appropriate for flap retraction, and the takeoff flaps are retracted. The aircraft is then accelerated to V_{ENR}, the single-engine en route climb speed. The third segment ends at the earlier of two events, either when the aircraft has achieved V_{ENR} or when 5 minutes have elapsed since the application of takeoff power. At the end of the third segment, the aircraft continues to climb to 1500 feet AGL at whatever speed was achieved in the third segment, but in no case less than flap retraction speed, and the power is reduced to maximum continuous power. The configuration during the third segment is:

- One engine windmilling
- One engine at takeoff thrust
- Flaps in the takeoff position

- Speed—V_2
- Accelerate to flap retraction speed
- Retract flaps
- Accelerate to V_{ENR}
- Upon achieving V_{ENR}, or after 5 minutes at takeoff power, reduce power to maximum continuous power
- Continue climb to 1500 feet AGL.

The minimum climb gradient that the aircraft must demonstrate in the final-segment climb to 1500 feet AGL is 1.2 percent for two-engine turbojets. When the aircraft reaches 1500 feet AGL, the takeoff climb is considered to be complete and the aircraft is considered to be en route. Figure 8-4 is a graphic representation of the FAR 25 takeoff-climb profile.

In addition to the maximum-takeoff-weight-permitted charts, manufacturers of FAR 25 aircraft are required to provide charts showing the actual gradients that can be expected to be realized during single-engine climb for the various segments. In order to use these charts for obstacle clearance, the FAA requires that the gross climb gradient be reduced by 0.8 percent during the takeoff phase and 1.1 percent during the en route climb phase; therefore, the minimum 2.4-percent gross climb gradient in the second segment will yield a 1.6-percent net climb gradient (2.4 – 0.8) to be used for obstacle clearance calculations.

LANDING

When the takeoff calculations and planning are done, many pilots believe they are finished with considerations of gross weight on aircraft performance. It is necessary to consult other charts for the landing phase of flight. There will also be a maximum landing weight permitted by climb requirements for the landing phase of flight. You might well wonder what climb requirements have to do with landing, and, of course, the answer is none if the approach and landing is successful.

Time and time again I have asked pilots the question "What is the termination point of an instrument approach procedure?" I get varying answers including DH, MDA, the missed-approach point, landing assured, actual touchdown, and being told to contact ground control. The termination point of a full instrument approach is *arrival at the missed-approach fix*. Just as every takeoff must be planned with the possibility of an engine failure in mind, each approach must be planned with the possibility of a missed approach in mind. If the missed approach is a possibility, then the aircraft must be capable of executing the missed-approach climb . . . on one engine. For this reason, there might be a restriction on the landing weight of the airplane.

Figure 8-5 is a landing-weight-limit chart similar to one used for a popular corporate jet. The chart shows the landing-weight limits as limited both by the approach-climb and by brake energy. The brake-energy limit is that weight at which there will be sufficient brake energy to stop the aircraft within the landing distance

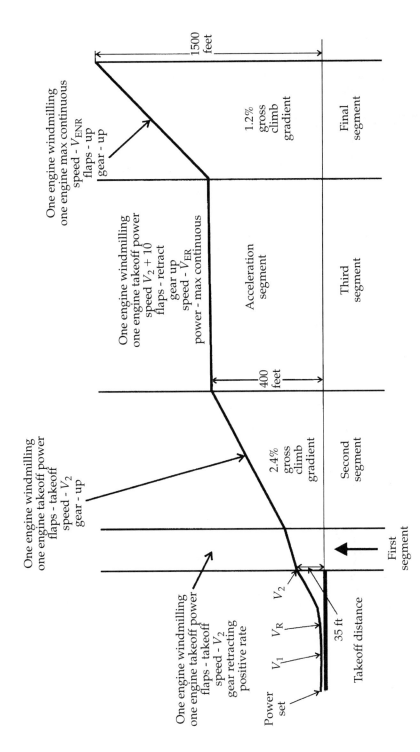

Fig. 8-4. The FAR-25 climb profile.

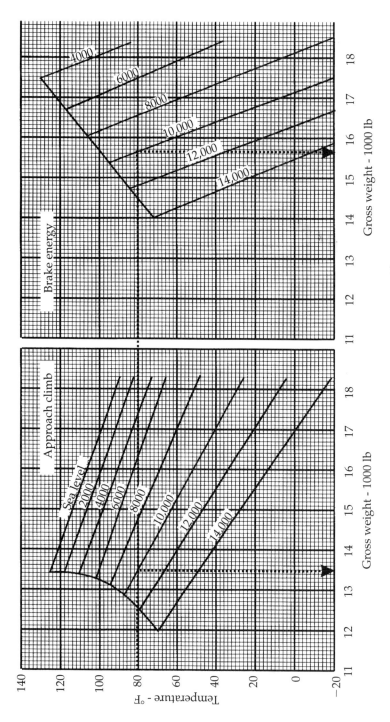

Fig. 8-5. This chart depicts the maximum landing weight permitted by climb requirements and brake-energy limits.

specified. The approach-climb limit ensures that the aircraft will achieve a minimum 2.1-percent gross climb gradient in the following configuration:

- One engine windmilling
- One engine at takeoff power
- Gear up
- Flaps—20°
- Speed—V_{APP}

The chart is entered on the left at the landing-field temperature and followed horizontally until intercepting the proper field elevation line, in this case 80°F and 10,000 feet. Directly below this intersection is the landing weight as limited by the approach climb: 13,400 pounds. Continuing the horizontal temperature line to the second part of the graph to intersect the second 10,000-foot altitude line and proceeding vertically below that intersection will give the landing-weight limit as limited by brake energy: 15,600 pounds. Because the maximum certified structural landing weight is 15,300 pounds for this particular aircraft, there will be sufficient brake energy to stop the aircraft in the landing-field length. Notice, however, that the approach-climb limitation will limit the landing weight to 13,400 pounds. This then will become the maximum allowable landing weight under the ambient conditions stated. (Again, wind and runway gradient have not been taken into account in this illustration.)

FAR 25 aircraft are required to meet one other climb requirement, the landing climb requirement. To be certified, the aircraft is required to demonstrate the ability to achieve a minimum gross climb gradient in the landing configuration of 3.2 percent. The landing-climb configuration is as follows:

- Two engines at takeoff power
- Gear down
- Landing flaps
- Speed V_{REF}

The landing-climb configuration is the only time FAR 25 addresses all engine performance in conjunction with minimum climb gradients.

A FLIGHT FROM VAN NUYS

Let's follow a flight from Van Nuys, California, to Las Vegas, Nevada, in a Cessna Citation 500. We will be departing on VNY Runway 16R. It's 75°F at VNY, wind 150 at 8 knots, and the altimeter is 30.03. Our aircraft, N501JC weighs 6972 pounds empty. We will be taking three passengers and a crew of two. With full fuel (3800 pounds), our takeoff weight will be 11,622 pounds (maximum 11,850), and our center of gravity will be just forward of the aft limit.[4]

The preflight has been completed and the before-start checklist is also complete. These items, along with copying the clearance, are typically completed before the passengers arrive. When the passengers arrive, they will be briefed on the safety items found in the aircraft, and we will depart.

We have received the following clearance:

"November 501 Juliet Charlie is cleared to the Las Vegas McCarran International Airport via the Newhall Five Departure, Daggett Transition, Creso Three Arrival. Maintain 4000 feet. Expect Flight Level 270 five minutes after departure. So Cal departure on 124.6. Squawk 4752."

Referring to Fig. 8-6, we find that the maximum takeoff weight permitted by climb requirements for our ambient conditions is the maximum structural weight of 11,850. We are below this weight, so we are assured of at least a 2.4-percent gross-climb gradient in the second-segment climb.

Refer to Fig. 8-7 for the appropriate power setting speeds for takeoff. Takeoff power will be 93.8 percent N_1. This is found by entering the table at the ambient temperature of 75°F. Project a line straight up to the first of the curve for the field elevation or the Maximum N_1 curve. In this case, we reach the maximum curve first and will go straight across to the left to find the takeoff N_1 of 93.8 percent. The same procedure is used with Fig. 8-8 to find the maximum continuous thrust setting of 92.6 percent N_1 and with Fig. 8-9 to determine the initial climb power of 94.0 percent N_1.

Table 8-2 will give us the V_1 of 102 KIAS, V_R of 102 KIAS, V_2 of 112 KIAS, V_{ENR} of 148 KIAS, and runway length required of 4000 feet for our takeoff. Runway 16R at VNY is 8001 feet long, so we have twice the amount the manufacturer requires.

Table 8-2. Takeoff distance V_1, V_2, V_R, and V_{ENR}
Flaps 15°
1,000 feet MSL
Gross weight—11,850

Temperature °F	V_1	V_R	V_2	V_{ENR}	Runway
−25	99	102	112	148	2,800
0	99	102	112	148	2,900
25	99	102	112	148	3,350
50	100	102	112	148	3,500
75	102	102	112	148	4,000
100	102	102	112	148	5,400

Notice the Newhall Five Departure shown in Fig. 8-10. The note in the lower left corner of the plan view indicates that this SID requires a minimum climb of 370 feet per nautical mile to 7000 feet MSL. Will we be able to make that gradient? Our flight manual information indicates that we will have a 3-percent net-climb gradient through 7000 feet. That doesn't sound like much. Let's see how many feet per nautical mile that will be. Our single-engine en route climb speed is 148 KIAS.

$$148 \text{ NM/hr.} \times 6000 \text{ ft/NM} = 888,000 \text{ ft/hr;}$$

$$888,000 \text{ ft/hr} \div 60 \text{ min/hr} = 14,800 \text{ ft/min horizontal speed.}$$

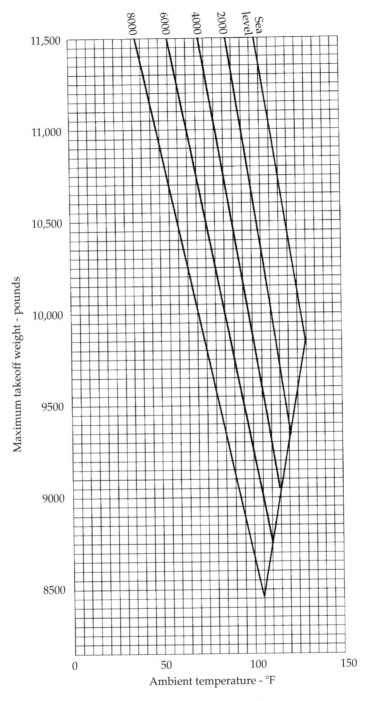

Fig. 8-6. This chart depicts the maximum takeoff weight allowed by climb requirements for the Cessna Citation CE-500 aircraft.

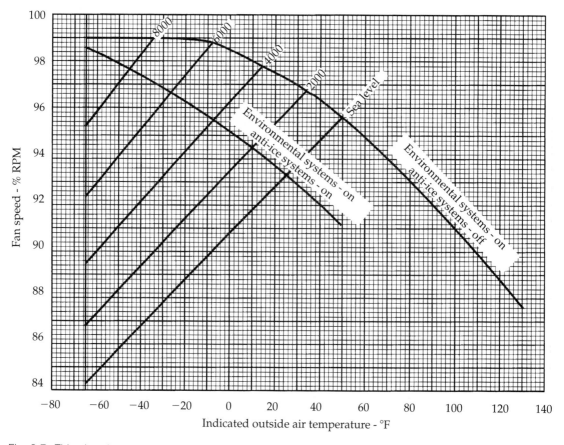

Fig. 8-7. This chart is consulted to determine the maximum power permitted for takeoff under ambient conditions.

Our net climb gradient is 3 percent. That is 1 foot up for each 100 feet horizontal.

$$14,800 \text{ ft/min} \times 0.03 = 444 \text{ ft/min rate of climb}$$

We have a problem. We cannot comply with the SID on one engine.

$$14,800 \text{ ft/min} \div 6000 \text{ ft/NM} = 2.46 \text{ NM/min}$$

$$444 \text{ ft/min} \div 2.46 \text{ NM/min} = 180.5 \text{ ft/NM}$$

If you look at a VFR chart for VNY, you will see what the problem is. Van Nuys lies in the San Fernando Valley (of *Valley Girl* fame). There are mountains between the Van Nuys Airport and the Palmdale VOR. Does this all mean that we can't go? No, but it sure would make you think twice in actual IFR conditions. The SID is an ATC clearance. If we lose an engine, we will be required to notify ATC of, among other things, our inability to comply with the SID.

We will look at one more chart before we decide we are legal to fly this trip. We will use approximately 1000 pounds of fuel on our flight to Las Vegas. That will make our landing weight 10,622 pounds. The chart for maximum landing weight

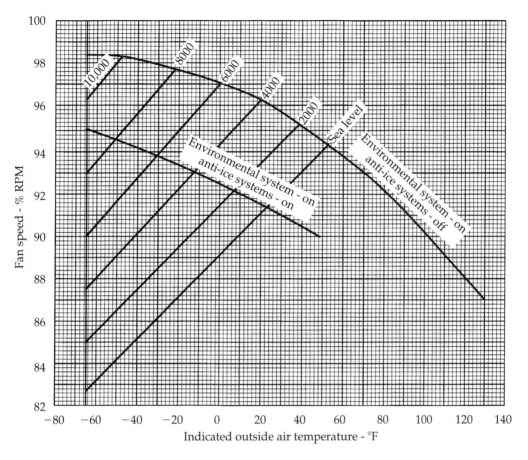

Fig. 8-8. This chart is consulted to determine the maximum continuous power permitted for the ambient conditions.

permitted by climb requirements (Fig. 8-11) indicates that we are legal up to our maximum structural weight of 11,350 pounds. (It's 90°F at LAS at a field elevation of 2177 feet.)

The passengers are loaded and briefed, and we are ready to start engines. The right engine is normally started first on this aircraft since it is the one opposite the door and since additional ground cooling is available with this engine running. Engine-start on a Citation is a semiautomatic procedure. The start button is depressed to engage the starter and start the rotation of the engine. When 8–10 percent N_2 is reached, the throttle is brought out of the idle-cutoff position to the idle-power position. This automatically activates the ignitors and introduces fuel to the engine. Within 10 seconds, the engine will light, and we will observe the ITT gauge to monitor the start. By 20 percent N_2, we should see an indication on the N_1 gauge, and oil pressure should be increasing. The start will automatically terminate at 32 percent N_2.

Because this start has been with the aid of an external power unit, we will not bring the generator online until the power unit is disconnected. When the right en-

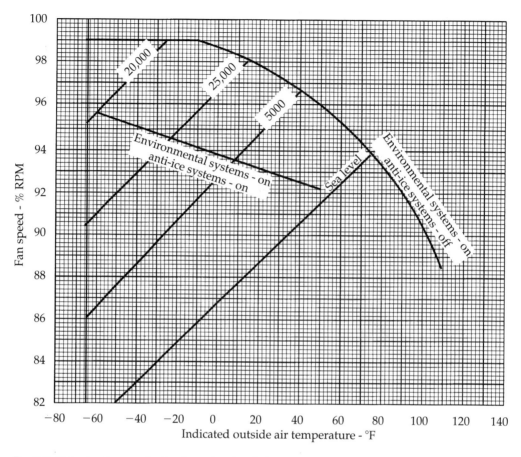

Fig. 8-9. This chart is consulted to determine the climb power permitted for the ambient conditions.

gine is stable, we will start the left. When the left engine is stable, the power cart will be disconnected and we will bring the generators on line.

Let's monitor the cockpit voice recorder from this point through landing at Las Vegas. (**Capt.** = captain; **F/O** = first officer; **ATC** = air traffic control.)

Capt.	Okay, let's have a taxi clearance and the taxi checklist, please.
F/O	Van Nuys Ground, Citation 501 Juliet Charlie at Million Air ready to taxi, we have information Tango and clearance.
ATC	One Juliet Charlie, taxi to runway 16R.
F/O	Brakes.
Capt.	Checked left.
F/O	Checked right. Flaps.
Capt.	Checked and set for takeoff. 15°.
F/O	Speed brakes.
Capt.	Checked and stowed, light out.
F/O	Takeoff trim.

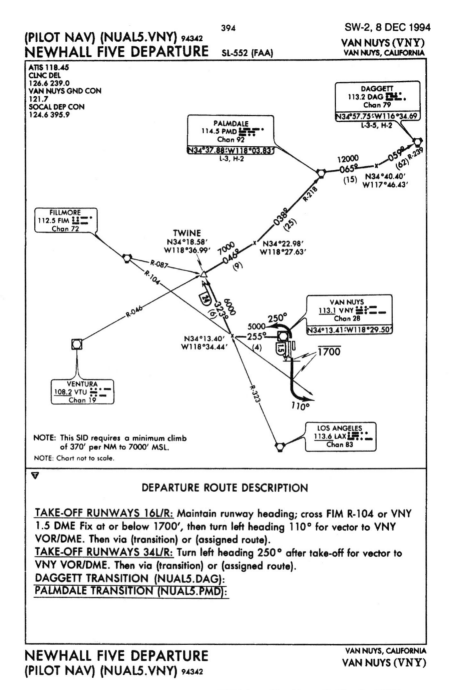

(PILOT NAV) (NUAL5.VNY) 94342
NEWHALL FIVE DEPARTURE SL-552 (FAA)

SW-2, 8 DEC 1994
VAN NUYS (VNY)
VAN NUYS, CALIFORNIA

ATIS 118.45
CLNC DEL
126.6 239.0
VAN NUYS GND CON
121.7
SOCAL DEP CON
124.6 395.9

DAGGETT
113.2 DAG
Chan 79
N34°57.75'W116°34.69'
L-3-5, H-2

PALMDALE
114.5 PMD
Chan 92
N34°37.88'W118°03.83'
L-3, H-2

12000
065°
(15)
059°
(62) R-239
N34°40.40'
W117°46.43'

R-218

FILLMORE
112.5 FIM
Chan 72

038°
(25)

TWINE
N34°18.58'
W118°36.99'

7000
046°
(9)

R-087

R-104

R-046

6000
323°
(6)

VAN NUYS
113.1 VNY
Chan 28
N34°13.41'W118°29.50'

250°
5000
255°
(4)

1700

N34°13.40'
W118°34.44'

VENTURA
108.2 VTU
Chan 19

R-323

110°

N34°22.98'
W118°27.63'

LOS ANGELES
113.6 LAX
Chan 83

NOTE: This SID requires a minimum climb
of 370' per NM to 7000' MSL.

NOTE: Chart not to scale.

DEPARTURE ROUTE DESCRIPTION

TAKE-OFF RUNWAYS 16L/R: Maintain runway heading; cross FIM R-104 or VNY
1.5 DME Fix at or below 1700', then turn left heading 110° for vector to VNY
VOR/DME. Then via (transition) or (assigned route).
TAKE-OFF RUNWAYS 34L/R: Turn left heading 250° after take-off for vector to
VNY VOR/DME. Then via (transition) or (assigned route).
DAGGETT TRANSITION (NUAL5.DAG):
PALMDALE TRANSITION (NUAL5.PMD):

NEWHALL FIVE DEPARTURE
(PILOT NAV) (NUAL5.VNY) 94342

VAN NUYS, CALIFORNIA
VAN NUYS (VNY)

Fig. 8-10. The Newhall Five Departure (SID) from Van Nuys, California (VNY).

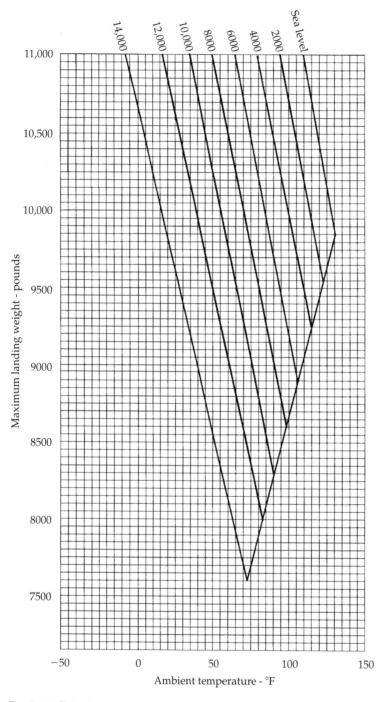

Fig. 8-11. This chart is consulted to determine the maximum landing weight permitted by climb requirements. This chart is limited by the climb requirements in the approach-climb configuration.

Capt. Three checked and set.

F/O Flight controls.

Capt. Checked full travel, top and bottom.

F/O Avionics and flight instruments.

Capt. Okay, flight instruments first. I have zero on the airspeed. Attitude gyro erect and no flags. Altimeter set 30.03 and 800 feet. Zero on the vertical speed. RMI and HSI heading check with the compass heading 070. No flags.

F/O Checked on the right also.

Capt. Radios. We have Van Nuys ground on the number one com, So Cal departure on number two. Nav one on Van Nuys and number two on LAX. The DME is tuned to Van Nuys.

F/O Autopilot.

Capt. Engaged and checked once with the disengage button, once with the trim. Flight director on, go-around mode engaged, heading set.

[The flight director is normally set to the go-around mode for takeoff. This will command a 7.5° noseup pitch initially. This pitch attitude will assure V_2 in the event of an engine failure. When it is determined that both engines are operating, the pitch attitude will be increased to 10–12°. This will give a good rate of climb, but will be a comfortable pitch attitude for passengers.]

F/O Checked. Altimeter and altimeter alert.

Capt. 30.03 on this side, 800 feet. Altitude alert set to 4000 feet.

F/O 30.03 and 780 on this side. Computations.

Capt. Let's see. You have V_1 of 102 set on your airspeed bug. 102 will also be rotation speed. I have V_2, 112 bugged on my side. Takeoff power is 93.8 set on the counter.

F/O Taxi check complete. Standing by on the before-takeoff check.

Capt. Okay. For a briefing, this will be an IFR departure in VFR conditions from runway 16R at Van Nuys. Runway required is 4000 feet, available 8001. We are cleared via the Newhall Five Departure to maintain 4000 feet expecting Flight Level 270 in five minutes. The departure calls for a straight climb to 1.5 DME at or below 1700 feet then a left turn to 110° for radar vectors. I will advance the power, I want you to set the power. Please call airspeed alive, 70 knots for a crosscheck. V_1 and rotate. Call positive rate, and I will command gear up. Call V_2 plus 10 and I will command flaps up. In the event of an engine failure, any red light on the panel, or loss of directional control prior to V_1 we will abort the takeoff. I will retard the throttles, apply brakes, and command speed brakes. If a problem occurs after V_1, we will continue the takeoff and consider it an airborne emergency. We will plan a VFR return to this runway. Please call 1000 before all assigned altitudes and alert me to any deviations in heading, altitude, or airspeed. Any questions?

F/O No.

Capt. Okay, let's get a takeoff clearance, and when we receive that, I'd like the before-takeoff checklist, please.

F/O   Van Nuys Tower, Citation One Juliet Charlie ready to go 16R at the north end, IFR.

ATC   Citation 501 Juliet Charlie, you are cleared for takeoff.

F/O   501 Juliet Charlie is cleared for takeoff. Pressurization, normal. Temperature select is in auto. Transponder and exterior lights are on. Ignition, pitot heat, and strobes on your side, please.

Capt. Ignition, pitot heat, and strobes on.

F/O   Engine instruments checked, annunciator panel clear.

Capt. Okay, all heading indicators check 160. Runway 16. Set power, please.

F/O   Power set, airspeed alive.

F/O   Seventy knots crosscheck.

Capt. Check.

F/O V_1.

[At V_1, the captain removes the hand from the throttles and places both hands on the yoke. The takeoff is now committed.]

F/O   Rotate.

[At rotation speed, the captain smoothly rotates the aircraft to 7.5°. As both engines have continued to function, the pitch attitude is increased to 10°. A pitch-synch button on the right yoke horn will synch the flight director command bars with the selected pitch attitude. The aircraft will be accelerated to 200 KIAS for the climb. If you remember our discussion of jet engines, you will remember that the engine becomes more efficient as the airspeed increases. This effect becomes noticeable at about 200 KIAS at sea level. Also, 200 KIAS is the speed at which compressibility affects become beneficial in increasing lift. When the aircraft is allowed to accelerate to 200 KIAS and then the pitch attitude is increased to maintain this speed in the climb, the performance will be optimized. The climb speed will be reduced 1 knot per thousand feet.]

F/O   Positive rate.

Capt. Gear up. Flaps up. Yaw damp on. After-takeoff check, please.

ATC.   One Juliet Charlie, turn left heading 110. Contact So Cal. Have a nice trip.

F/O   So Cal now, g'day, One Juliet Charlie. . . . So Cal. Citation 501 Juliet Charlie with you out of two thousand for four thousand.

ATC   Roger, One Juliet Charlie. Farther left now 340, vectors for Palmdale. Climb and maintain one zero thousand.

F/O   Roger, So Cal. Left 340, up to one zero thousand. One Juliet Charlie.

Capt. OAT 60°F. Climb power for 5000 please.

[Climb power is checked and adjusted as necessary every 5000 feet during a climb in this aircraft.]

F/O   Landing gear up and indicating, flaps up, landing lights off, yaw damp on, ignition normal when you get a chance, climb power is set, pressurization checks. After-takeoff check complete. Would you like Palmdale on your side now?

Capt. Please.

F/O   Nine for ten. Seatbelt light off.

Capt.	Roger. See if we can get higher now.
ATC	One Juliet Charlie, now maintain one four thousand. Contact Joshua on 126.1.
F/O	Up to one four thousand and Joshua on 126.1. Good day, now, One Juliet Charlie.
F/O	Joshua Approach, Citation 501 Juliet Charlie, out of one-one for one four thousand.
ATC	Roger One Juliet Charlie. Climb now to one seven thousand, intercept the Palmdale 218 radial. Palmdale altimeter 30.01.
F/O	Up to one seven thousand, intercept the Palmdale 218. One Juliet Charlie.
Capt.	Okay, 038° to Palmdale and needle alive. Tune and identify Daggett on your side, please.
F/O	Daggett on the number two.
ATC	Citation One Juliet Charlie. Now maintain Flight Level 270. Contact Los Angeles Center 133.55.
F/O	Up to two seven zero and Los Angeles on 133.55. Good day, one Juliet Charlie.
F/O	Flight level 180. Altimeters 29.92. Air conditioning off. Los Angeles, this is Citation 501 Juliet Charlie with you out of one nine zero for two seven zero.
ATC	Roger One Juliet Charlie.
F/O	Flight level two five zero. Oxygen set to 100 percent.
Capt.	O-two at one hundred. Thank you.

[Above Flight Level 250, the oxygen masks must be stowed in the quick donning position and the selector must be set to 100-percent oxygen.]

F/O	Two six zero for two seven zero.
Capt.	Roger.
Capt.	Level two seven zero. Autopilot on. Cruise check, please.

[It is not mandatory that the Citation be flown on autopilot at altitude, in fact, the airplane is very stable and easy to hand fly. The autopilot will, however, give a better ride to the passenger than the best pilot. Also, a truly professional pilot will use all aids available to make the flight safer and more efficient. Letting the autopilot fly at altitude allows the pilot to "rest up" for an instrument approach. Even with the autopilot engaged, however, one pilot—in this case, the captain—must be in command of the aircraft and closely monitoring all parameters.]

F/O	Cruise power set. Pressurization checked. Cruise check complete.
ATC	Citation One Juliet Charlie, cross Jokur at Flight Level 240, Saras at fourteen, Whigg at twelve. Daggett altimeter 30.11.
F/O	Roger, LA. Jokur at 240, Saras at 14. Whigg at 12. One Juliet Charlie.

[This is a clearance we expected. Passing Daggett, we will be flying the Creso Three Arrival to Las Vegas (Fig. 8-12). Our current ground speed is 320 knots. We'll plan the descent for a ground speed of 360 knots, or 6 miles per minute. Our first fix is 10 miles east of Daggett, and we need to loose 3000 feet before we get there. To calculate the point at which we need to start our descent, we'll multiply the al-

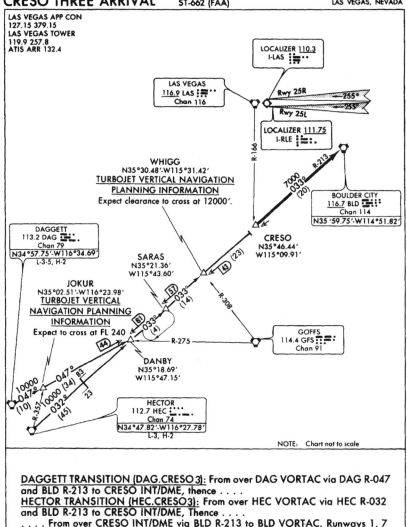

(CRESO.CRESO3) 94286
CRESO THREE ARRIVAL   ST-662 (FAA)

McCARRAN INTL
LAS VEGAS, NEVADA

LAS VEGAS APP CON
127.15 379.15
LAS VEGAS TOWER
119.9 257.8
ATIS ARR 132.4

LOCALIZER 110.3
I-LAS

LAS VEGAS
116.9 LAS
Chan 116

Rwy 25R   255°
Rwy 25L   255°

LOCALIZER 111.75
I-RLE

WHIGG
N35°30.48'-W115°31.42'
**TURBOJET VERTICAL NAVIGATION
PLANNING INFORMATION**
Expect clearance to cross at 12000'.

R-213

7000
033°
(20)

BOULDER CITY
116.7 BLD
Chan 114
N35 '59.75'-W114°51.82'

DAGGETT
113.2 DAG
Chan 79
N34°57.75'-W116°34.69'
L-3-5, H-2

CRESO
N35°46.44'
W115°09.91'

(23)

SARAS
N35°21.36'
W115°43.60'

(43)

JOKUR
N35°02.51'-W116°23.98'
**TURBOJET VERTICAL
NAVIGATION PLANNING
INFORMATION**
Expect to cross at FL 240

(57)   033°
(14)

R-308

(81)   033°
(4)

(44)   R-275

GOFFS
114.4 GFS
Chan 91

DANBY
N35°18.69'
W115°47.15'

10000   047°
047°   10000 (34)   83
R-351   032°   23
(10)   (45)

HECTOR
112.7 HEC
Chan 74
N34°47.82'-W116°27.78'
L-3, H-2

NOTE: Chart not to scale

DAGGETT TRANSITION (DAG.CRESO 3): From over DAG VORTAC via DAG R-047
and BLD R-213 to CRESO INT/DME, thence . . . .
HECTOR TRANSITION (HEC.CRESO3): From over HEC VORTAC via HEC R-032
and BLD R-213 to CRESO INT/DME, Thence . . . .
. . . . From over CRESO INT/DME via BLD R-213 to BLD VORTAC. Runways 1, 7
and 19 expect vectors to final approach course. From BLD VORTAC, expect ILS
approach Runway 25L.

CRESO THREE ARRIVAL
(CRESO.CRESO3) 94286

LAS VEGAS, NEVADA
McCARRAN INTL

Fig. 8-12. The Creso Three Arrival (STAR) to McCarran International Airport in Las
Vegas, Nevada (LAS).

titude change in thousands—3—by three and start there. That means no later than 9 miles out. The Daggett VOR is as good a place as any to begin because it is 10 miles from the fix. We'll descend at 2000 per minute. Six miles per minute will mean we will cover 9 miles in 1 minute and 30 seconds. At 2000 feet per minute, we will descend 3000 feet in that time. Power will be adjusted to keep the speed below the redline. The weather is smooth, and there are no reports of turbulence in the desert today. During the summer, however, it gets pretty bumpy at low altitudes over the Mojave Desert, and we would want to adjust our speed accordingly. Descending into a turbulent air layer at a high rate of speed will at the very least wake everyone up!]

F/O   Crossing Daggett now.

Capt.   Roger, set 047 as the outbound course. Please tune and identify Boulder City on the number two and set 033 inbound. After we cross DANBY, the DAG 44 DME fix, we'll switch the DME to BLD. Out of two seven zero for two four zero.

F/O   Boulder City tuned and identified on number two.

[This is a good time, if you haven't already done so, to determine if your passengers will be needing a cab at the destination. Make these unicom calls before you get switched to approach and everyone up-front gets busy.]

F/O   Out of two five zero for two four zero.

Capt.   Roger. Please call the Daggett 18 DME.

[We need to cross SARAS at 14,000. This fix is 38 miles from JOKUR. Using our previous rule of thumb, we will need to begin the descent when we are no closer than 30 miles to the fix. When the DME from DAG reads 18, we will begin down at 2000 fpm.]

F/O   Approaching the 18 DME fix.

Capt.   Roger. Out of 240 for 140. Autopilot off. Descent checklist please. Call the 44 DME fix for DANBY, please.

F/O   Okay. Defog fan on, foot warmers closed, pressurization set. Altimeters to go at 18,000. Monitor ATC for me, please. I'm going off the frequency to get ATIS.

Capt.   Roger, monitoring.

F/O   Back on frequency. Las Vegas information Whiskey. Clear and forever. Wind 250 at 15, gusting 20. Temperature's 90. Altimeter 30.12. Arriving runway is 25.

Capt.   Roger.

F/O   44 DME.

Capt.   Roger. Turning left to intercept the BLD 033. Set BLD on number one, please.

F/O   Boulder City tuned and identified on number one. Inbound course of 033 set. Out of 15 for 14.

Capt.   Roger.

ATC   Five Zero One Juliet Charlie, Las Vegas Approach now on one two seven point one five.

F/O   Las Vegas on one twenty seven fifteen. Five oh One Juliet Charlie.

F/O	Las Vegas, Citation 501 Juliet Charlie with you at one four thousand with Whiskey.
ATC	Good afternoon One Juliet Charlie. Descend to and maintain 8000. Las Vegas altimeter 30.11. Direct Boulder City for now. Expect vectors ILS 25L.
Capt.	Tell them we're familiar with the area and we'd like to cross Henderson for the number of 25 on a downwind for 19L. Out of 14 for 8. Seatbelt light on, please.
F/O	Roger. Las Vegas, One Juliet Charlie is familiar and would like to cross Henderson for the number 25 downwind for 19L.

[The weather in Vegas is good VFR. No ceiling and visibility unlimited. Vectors for an ILS will take us at least 20 miles east of the airport and put us in a lineup of airliners. Additionally, if we land on either 25 Left or Right runways, we will have a long taxi to our FBO and a long wait to cross Runway 19L. Henderson is the Henderson Sky Harbor Airport about 7 miles south of McCarran.]

ATC	Roger One Juliet Charlie. Direct Sky Harbor. Cross the numbers 25 at four thousand. Heading 010 over the numbers downwind for 19L. You are cleared for the visual approach Runway 19L.
F/O	Nine for eight.
Capt.	Roger. This will be a visual approach to 19L. Please monitor the approach and call approaching all assigned altitudes. When we turn final, please monitor airspeed, altitude, and rate of descent. Call airspeeds in relation to ref and altitudes above ground. Alert me of any sink rate greater than 1000 feet per minute above 1000 AGL and any rate above 500 feet per minute below that. Before-landing check, please.
F/O	Roger. Avionics and flight instruments. I have the Las Vegas VOR on both navs. DME on LAS. Altimeters set and crosschecked. Radio altimeter set to 400 AGL. Computations, V_{REF} will be 108 bugged on both airspeed indicators, go around N_1 will be 92.0 percent. Holding at flaps approach.
Capt.	Roger.
F/O	Crossing Henderson now.
Capt.	Roger. Flaps approach, please. Out of 8 for 4.
F/O	Speed checks. Flaps selected and indicating approach. Holding checklist at landing gear.
F/O	Five for four.
Capt.	Roger.
Capt.	Crossing the numbers for 25. Heading 010.
ATC	One Juliet Charlie, contact tower now, 119.9.
F/O	Tower on nineteen nine. Good day.
F/O	Las Vegas Tower, Citation 501 Juliet Charlie with you downwind on the visual for 19L.
ATC	One Juliet Charlie, good afternoon. You are cleared to land runway 19L. Wind 230 at 20.
F/O	One Juliet Charlie, cleared to land 19L.

Capt.	Gear down, continue checklist please.
F/O	Speed checks, gear selected down, gear down and locked. Three green. Exterior lights on, ignition on, annunciator panel clear, holding on full flaps and yaw damper.
Capt.	Roger. Call 400 feet.
Capt.	Turning final now.
F/O	Ref plus ten, sink 600. 400 feet AGL now.
Capt.	Full flaps, yaw damp off. Call 50 feet.
F/O	Flaps to full, yaw damp off, ref plus 5, sink 5.
F/O	Fifty feet. At ref.
Capt.	Power coming to idle.

[At touchdown, the captain will tap the brakes to make sure they are working, then will not apply brakes again until the aircraft has slowed significantly.]

Capt.	Extend speed brakes, call 60 knots.
F/O	Sixty knots.
ATC	One Juliet Charlie. Turn right when able. Ground point nine.
F/O	Las Vegas ground, Citation One Juliet Charlie with you off 19L to Signature.
ATC	Roger, one Juliet Charlie, taxi to Signature.
Capt.	After-landing check, please.
F/O	Flaps up and indicating, speed brakes retracted, ignition, pitot heat and strobes when you get a chance. Transponder to standby. Landing lights off.
Capt.	Ignition to normal, pitot heat and strobes off.
Capt.	Shutdown checklist please.
F/O	Brakes set, ac power and master avionics off, overhead and defog fans off, generators off, throttles idle cutoff, passenger-advisory lights off, exterior lights off, battery off.

We could follow the flight back to Van Nuys, but it would be much the same as the flight to LAS. Let us instead address some of the differences between a flight in a jet and a flight in a cabin-class piston twin.

JET DIFFERENCES

Power settings Power settings for this aircraft, since it is a turbofan, are made in reference to percent N1. Other jets might use EPR, but whichever is used, there is only one gauge to reference when setting power. The power charts for takeoff power will be referenced to field elevation and ambient air temperature. The temperature is the most limiting parameter, however. Close inspection of the takeoff thrust setting chart for this aircraft (Fig. 8-7) will show that altitude becomes limiting only at very low temperatures. Also notice from this chart the reduction in takeoff thrust that must be made with engine anti-ice on. This was discussed in chapter 4.

Jets cruise at maximum power. The jet engine is most efficient when operated at its full-design RPM. Remember, the top 30 percent of the power in a turbofan engine is generated in the top 10 percent of rotation, and this is not a linear curve, but an exponential one so that any RPM reduction on the top is a great loss of power. The amount of money saved in fuel consumption by cruising at long-range cruise in a jet is more than lost by the expense of additional hours accumulated on the engines and airframe, and if a headwind of more than 25 percent of cruise TAS exists, no fuel savings is likely to be realized by using long-range cruise instead of maximum cruise. The most economical altitude for a jet to cruise is roughly at the tropopause (about 36,086 in an ISA condition) where the temperature stabilizes.

Only maximum power settings for takeoff, maximum continuous power, climb, and cruise are presented in FAR 25 flight manuals. Power settings used for maneuvers, air work, and approaches are passed on from instructor to student. Many jet pilots prefer to use fuel flow for these power settings since fuel flow in a turbine aircraft is an extremely accurate measure of power output.

Lack of propellers This might seem obvious, but the problems that are eliminated when propellers are eliminated are important. First and foremost, the maintenance associated with propellers is not needed. Propellers are complex pieces of machinery with springs, pressure chambers, nitrogen charges, counterweights, governors, and all manner of stuff that just loves to get out of adjustment and needs the ministrations of a mechanic. Props usually have a TBO much less than the engine they are attached to. Also, propellers, like any object rotated at high speeds, must be kept in delicate balance. This is true with jet engines, of course, but the propeller, unlike the turbine wheel, is constantly exposed to the elements and acquires much more damage from water (yes, water is VERY hard on props), rocks, and other manner of debris. Propellers are also a liability in the event of an engine failure. Not only must the windmilling propeller be feathered and put in the minimum-drag configuration to even have the possibility of continued flight, but the loss of the propeller slip stream over the affected wing results in a loss of lift on that wing. Jet engines cause no such problems.

There is no P-factor and no left-turning tendency during takeoff and climb with a jet aircraft. And of course, all the noise and vibration associated with propellers and their engines is absent in a jet aircraft.

Propellers do have some advantages that are missing in jet aircraft. I remember making a statement something to the effect that I would never miss propellers again after I had my first experiences in jet airplanes. That opinion lasted until the first time I was on a slick, wet runway with a jet. I certainly missed the instant power response of piston engines and the beta- and reverse-thrust capabilities of turboprops. Jets are skittish little critters on slick pavement. Jets also have more tendency to weathervane into the wind during crosswind operations than do propeller-driven craft. Propellers provide a large gyroscopic stabilizing force. Propellers also act like huge rotating speed brakes on the runway, and I don't think any pilot is aware of how much props decelerate an aircraft until they aren't available.[5]

MANEUVERS
Takeoffs

Takeoffs are the forgotten maneuver. Once a pilot achieves a private certificate, it seems that all flight instructors assume that person now knows how to execute a takeoff, and instruction in that area ceases. Every pilot has been performing takeoffs of one type or another since the beginning of his or her flying career, and it is assumed that he or she knows all there is to know about this subject. That might well be true in some cases, but the true professional pilot will admit there is always more to learn. Technique can always be polished.

The takeoff, while probably the first maneuver the new pilot masters to his or her instructor's satisfaction, can also be the potentially most dangerous part of the flight. At no other point in the flight will the aircraft be as heavy as it is during takeoff nor will it again contain as much flammable fuel. All power available is already in use, leaving no margin for overcoming any unanticipated wind shear or other gust.

Another fact, appreciated by taildragger pilots, but vastly unacknowledged by today's "nosedraggers," is that the airplane is not a ground vehicle and is totally out of its element while on the ground. It handles very well and is nimble in its element in flight and has acceptable handling characteristics at taxi speeds. In the transition from taxi speeds to flight, however, the aircraft becomes a fractious and temperamental beast that has the potential to get away from the unwary. This is the transitional period, from being groundborne to being airborne, during which a rejected takeoff must be accomplished if a malfunction occurs.

Normal takeoffs The entire length of the runway should always be used for takeoff. ATC might urge an intersection takeoff to expedite traffic, but I have never found that concern to override the safety concerns of having as much pavement as available in front of the aircraft when the throttles are pushed up. Line the aircraft up on the centerline of the runway and make sure the nosewheel is straight.

The pilot flying will advance the throttles and command the nonflying pilot to set the power. The pilot flying will not relinquish the throttles, but will loosen his or her grip to allow the throttles to be advanced and trimmed. Power should be advanced slowly and smoothly until the engines have stabilized and symmetrical thrust is assured. The throttles can then be smoothly advanced to the takeoff-power setting.

A slight forward pressure should be kept on the control column to keep the nosewheel firmly on the runway. In an aircraft with nosewheel-tiller steering, the captain should monitor steering closely until the rudder is effective.

After passing V_1, the last point to abort the takeoff has passed, and the pilot flying will place both hands on the yoke in preparation for rotation. At V_R, the aircraft should be smoothly and decisively rotated to the takeoff attitude. I'm sure you have seen jet takeoffs during which the nose is almost snapped into the takeoff attitude. This is not necessary and smacks of show-off flying. The rotation must be positive and at the proper speed. Early rotation will delay the takeoff. Late rotation will delay the takeoff, in fact, rotating at a higher speed will most likely require more back pressure on the column. The faster a wing at a negative angle of attack moves, the more downward lift it produces. It wants to now stick to the runway. Also, the rotation is liable to be more abrupt at a higher speed and upsetting to passengers. A smooth, positive rotation will result in a liftoff at a deck angle of about 7°–9°. Continue the rotation to the desired climb angle.

Landing-gear retraction should be accomplished as soon as the aircraft has attained a positive rate of climb as verified by two of the following sources: the altimeter increasing, the VSI increasing, or a positive rate as noted out the window. Flaps/slats are retracted according to the flap-retraction schedule specified by the manufacturer. Acceleration through V_2 will be rapid and no attempt should be made to maintain V_2 with both engines functioning. This is an emergency speed to be used only in the event of an engine failure, and attempting to maintain this speed with both powerplants working will result in an extremely high deck angle that will be upsetting to both passengers and pilot examiners (Fig. 8-13).

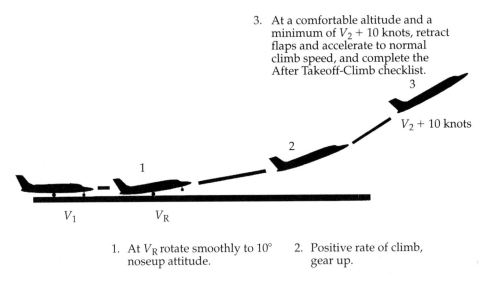

3. At a comfortable altitude and a minimum of V_2 + 10 knots, retract flaps and accelerate to normal climb speed, and complete the After Takeoff-Climb checklist.

3

V_2 + 10 knots

2

1

V_1 V_R

1. At V_R rotate smoothly to 10° noseup attitude.   2. Positive rate of climb, gear up.

Fig. 8-13. A normal takeoff in the Cessna Citation CE-500 is accomplished according to these procedures.

One-engine-inoperative takeoff For all practical purposes, you should consider a takeoff mandatory when an engine failure occurs above V_1 unless at least three times the required runway is available. This is especially true on anything but a level, dry runway. On hot days, the accelerate-go distance might well greatly exceed the accelerate-stop distance, and in this case, a rejected takeoff *may* be considered with plenty of excess runway available. NEVER ATTEMPT AN ABORT AFTER INITIATING ROTATION. If the nose is off the runway, fly the airplane. It is developing lift at this point, and you have no guidance as to how much runway will be required to land and stop the aircraft from this point.

The airplane will yaw toward the failed engine. Use whatever rudder pressure is required to maintain directional control, and use aileron to maintain a wings-level attitude. Most new jet pilots are surprised regarding what little rudder it takes to maintain heading during an engine-out takeoff in an airplane with aft-mounted engines. The amount of rudder for a jet is certainly much, much less than is required for a V_1 cut in the likes of a DC-3.

Rotation should be accomplished in the same manner as during a normal takeoff. At liftoff, rudder and aileron must be used properly and with discretion to maintain directional control and a wings-level attitude. Avoid undue rolling and yawing motions and overcontrolling. Any overcontrolling might induce a Dutch roll, which is highly undesirable in this phase of flight.

Maintain a flight attitude that will produce V_2, and climb to a minimum of 400 feet. The only events that will occur prior to 400 feet are acknowledgment of the problem and the pilot's command for (and subsequent) gear retraction. Remember, your most important job is to fly the airplane. The pilot not flying will receive your commands and perform any actions.

At 400 feet, command and verify that all memory immediate-action items on the engine failure checklist have been accomplished. Do not hurry. An engine loss in a jet is no real emergency unless you make it into one. Take your time and make sure things are done in an orderly manner. After the immediate-action items have been accomplished, accelerate the aircraft to single-engine climb speed and take no further action until a comfortable obstacle-clearance altitude has been reached. At this point, command the rest of the engine-failure checklist be accomplished.

Now a decision must be made about where to land the aircraft. Of course, if it is VFR you will turn downwind and command the pilot not flying to contact the tower and tell ATC you have had an engine failure and are returning to land. You might need to dump fuel to lighten the aircraft for landing; if so, you will need a clearance to a fuel dumping area.[6] In IFR conditions, you will need an approach clearance and vectors. If the departure airport is below landing minimums, you will need a clearance to your takeoff alternate. No matter which situation is pertinent, it should be noted that the last thing that is done is communication with ATC; when that communication is accomplished, it will be to inform a controller about the assistance you need to complete your flight safely. Too many pilots seem to need the security blanket of talking to ATC. Get the situation handled first. Get your aircraft configured for the flight and all checklists complete. Then and only then should consideration be given to talking to ATC (Fig. 8-14).

Rejected takeoff This is a maneuver that is best practiced in a good simulator. A high-speed abort is one of the more dangerous maneuvers a pilot can be asked to perform. The procedures specified for rejecting a takeoff in most jets are different from those required on a normal landing roll. The rejected takeoff is accomplished, according to regulation, by applying maximum braking, then retarding the throttles to idle, and deploying speed brakes or ground spoilers. A normal landing roll requires idle power and speed brakes before wheel brakes are applied. It is difficult to perform these items in a different order when the landing order of progression has been reinforced so many times.

5. Complete the After Takeoff-Climb and Engine Failure checklists.

4. Accelerate to V_{ENR}, reduce power to maximum continuous and climb to 1500 feet AGL.

3. Gear up when positive rate of climb is established. Maintain V_2 until 400 feet AGL or clear of obstacles, whichever is higher; accelerate to $V_2 + 10$ knots, and retract the flaps.

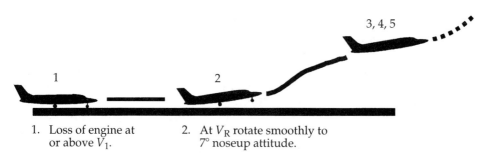

1. Loss of engine at or above V_1.

2. At V_R rotate smoothly to 7° noseup attitude.

Fig. 8-14. A takeoff continued after an engine failure in the Cessna Citation CE-500 requires adherence to the FAR-25 takeoff profile.

Directional control is a prime consideration during a rejected takeoff. Maintain the runway centerline while decelerating the airplane. Asymmetrical thrust due to an engine failure or directional control problems due to blown tires will try to pull you off to one side or another, but make every effort to keep the airplane straight and on the centerline. This gives you the most margin for error. If a skid does develop, try to overcome it with aerodynamic controls. Use rudder pressure to deflect the nose in the direction of the skid. Do not use reverse thrust when trying to overcome a skid.

Steep turns

Steep turns, those with a bank angle of more than 30°, are a maneuver included in the jet-training syllabus to assess the pilot's ability to maneuver the aircraft at high bank angles with precise control of heading, altitude, and airspeed. Obviously, executing this maneuver in the traffic pattern with passengers would be unwise, even though the airplane would have a safe margin above stall at the proper maneuvering speed.[7] Also, since a 45° bank, the bank angle used for most steep turns, imposes a 1.5-g load on the structure, a Mach buffet is likely to be the result of rolling into a 45° bank at high altitude near the limiting Mach number.

The speed at which steep turns are practiced varies with aircraft, but are usually in the 200- to 250-knot range. Your particular training syllabus will specify the speed. The Citation uses 200 KIAS, the DC-9 uses 220 KIAS, and the Sabreliner uses 250 KIAS. Whatever the speed specified, you will want to make sure that you have the power set and stabilized and the aircraft trimmed for this speed prior to commencing this maneuver.

The most common errors observed in steep turns are: improper rate of roll entering the turn (usually too slow), poor altitude control on turn entry and when established (usually a climb followed by a descent), poor bank control during the turn (usually shallowing the bank), poor speed control in the turn (usually too slow), and premature rollout from the turn. A good instrument scan and knowing the aerodynamics of the turn can go a long way toward avoiding these errors.

The first error, improper rate of roll entering the turn, is usually a result of the student's desire to be smooth with the aircraft. That is admirable, but the pilot must also be in command, and this maneuver requires a slightly higher than normal roll rate for entry. As a rule of thumb, the aircraft should have turned one-third of the angle of bank, or 15°, by the time the bank angle is established. This entry should be executed with smooth, steady control pressures.

The second common error, poor altitude control, is actually caused by a combination of effects. First of all, the pilot will be aware of the need to increase the angle of attack during the turn by applying back pressure to maintain altitude. You must not only compensate for the loss in vertical lift from the angle of bank, but you know you must also increase the angle of attack to compensate for the additional load factor caused by the turn. This increased lift will also result in increased drag, which will have an adverse effect on airspeed.

The first tendency of the aircraft as you roll into the turn is to climb. This is caused by a premature increase in the back pressure on the column plus the airplane's tendency to climb upon entering the turn. This aileron-induced effect is rarely taken into consideration by students because it is rarely explained by instructors.

To make a right turn, you apply right aileron. This causes the right aileron to rise into the airflow over the wing. This causes the airflow over the top of the right wing to be slowed and its pressure to increase slightly. Meanwhile, the airflow over the bottom of the wing is speeding up slightly, and its pressure is decreasing. This all causes a slight loss of lift on the right wing, and the wing will go down.

The left aileron is deflected downward at the same time. This has an effect on the left wing opposite that on the right. The air over the top of the wing will speed up, and the pressure will decrease. The air over the bottom of the wing will be slowed, and its pressure will increase. This will cause the left wing to create more lift, and it will rise. The lifting effect on the left wing, however, will be of a much greater magnitude than the loss of lift on the right wing. The wing with the aileron down tends to climb at a faster rate than the wing with the aileron up tends to descend. The net effect is that the airplane tends to climb initially. Also, the left wing, being on the outside of the turn, will be going slightly faster than the right wing. Its greater lift as it climbs in the beginning of the turn will cause the airplane to climb until the loss of vertical lift begins. This loss of lift does not usually begin until the angle of bank exceeds 15°.

Here is what happens. The hapless pilot begins the steep turn by rolling the aircraft in the desired direction and applying slight back pressure to compensate for the loss of vertical lift and increase in drag. This causes a climbing tendency. The aileron effect also causes the airplane to climb. The pilot does not notice this yet because all concentration is on establishing exactly 45° of bank. When the pilot does look, 50 to 100 feet has appeared on the altimeter! Now the correction begins. Just about the time the back pressure should be increased, the pilot exerts forward pressure to return to the target altitude. Before long, the altitude is 100 feet below the desired altitude. It is not uncommon for pilots to porpoise through 180° of the 360° turn before this effect is dampened out.

How can this be avoided? First of all, be prepared. Roll smoothly into the bank, applying slight forward pressure on the yoke until rolling through approximately 20°, then check the altimeter and use a steady back pressure to peg the altimeter on the desired altitude. Additionally, to maintain the desired airspeed, it will be necessary to increase power when rolling through about 30° of bank. Advance the power, or command that it be adjusted, approximately 0.05 EPR or about 3 percent N_1. (About 50 pounds per side fuel flow, if that's your preference.) Stop the bank at 45°, and keep it there. Varying the bank angle varies the vertical lift, and consequently the drag, and makes altitude control extremely difficult.

Control the altitude with elevator and strive to use steady control pressure rather than large movements. One degree of pitch change is a lot in a jet, and you should strive to keep your manipulations of the controls within that parameter. The attitude-direction indicator will show a 3° noseup pitch, dependent upon the aircraft you are flying and the airspeed selected for the turn.

The airspeed indicator is primary for deciding how much power you need. You will have to control your speed within ±10 KIAS, and this should also be done with small corrections. Because of the spoolup time inherent with jets, it is very easy to quickly get 180° out of phase with the power requirements. The best way to avoid this is to have a target power setting and vary it by small increments as soon as the airspeed indicator shows any tendency to move.

As soon as the bank is established, look at the HSI to determine how many degrees you turned on the roll-in. This is the number of degrees by which you will lead your roll-out. This should be about 15° if your roll was near the proper rate. Using this technique, you should be able to roll wings level within 5° of the desired heading.

The final error pilots make in steep turns is in the roll-out. Remember all that lift that was lost on the roll-in? Well, its going to come back when the wings are rolled level. You must be prepared to nail the altimeter by applying forward pressure on the yoke as the wings are leveled. The decrease in the angle of attack, of course, causes a decrease in drag, and so the power must be smoothly reduced by the amount it was increased, or the airspeed will increase. I have seen more than one beautifully executed steep turn ruined at the last minute by a balloon above the target altitude caused by the pilot's neglecting to apply the forward pressure to maintain altitude.

Think through this maneuver and all maneuvers before you perform them. Take your time to ensure that the aircraft is stabilized and trimmed for the speed you want to maintain during the maneuver. Make sure your heading, altitude, and speed are exact before you begin the maneuver. Be aware of the aerodynamics involved, and apply all corrections with smooth pressure in small increments. Instrument flying is nothing but a long series of corrections back to a desired heading, altitude, and speed. The smaller those corrections, the smoother the flight.

Stalls

Full stalls are not practiced or demonstrated in jet flight training. The approach to the stall is geared to teach the pilot to recognize the approach and take corrective measures before the stall develops. A properly trained pilot should never get his or her aircraft into the stalled condition and should be able to apply those corrective measures well before the stall occurs.

Stalls for the type-rating checkride are done in the clean, takeoff-flap, and full landing configuration. Precise control of altitude, heading, and airspeed are important in this maneuver, as in all others. Prior to beginning the maneuver, some preparation is necessary. Of course, these procedures will be governed by the flight manual for your particular aircraft, but in general, you will need to compute a landing reference speed for your gross weight, turn the ignition to ON or to CONTINUOUS, check the stall warning and/or stick shaker systems for proper operation, and compute the takeoff power setting for the ambient temperature.

Before initiating the stall series, turns should be done to clear the area of other traffic. There should not be a great deal of traffic, especially single-engine aircraft since stall practice should be accomplished at 10,000 to 14,000 feet MSL. This will keep you pretty much clear of other traffic, but you will be low enough not to contend with Mach effects. Make sure you thoroughly understand the recovery sequence for your particular aircraft.

All stalls for the type-rating ride will include recovery at the first sign of the stall. This will be different in different aircraft. It might be an aural stall warning, a stick shaker, an angle-of-attack indication, or, as in the case of the Citation 500 series, an aerodynamic buffet. No matter what the indication for recovery is, be sure you understand it and recognize its arrival.

Because so many jet pilots begin their careers in the Citation, let's see how the stall series will go in that aircraft. We will use the angle-of-attack indicator as a stall warning indicator. It has a white hatched area, known as *the box*, just before the stall, and we will use this for our recovery criteria. The aircraft will be prepared with ignition on, V_{REF} bugged on both airspeed indicators, and the yaw damper off. (This will not be true for all aircraft.) Takeoff thrust for ambient conditions will be computed and entered on the counter. Please take note: *These procedures will not hold true for all jets. The aircraft flight manual for the airplane involved MUST be consulted for proper procedures.*

Clearing turns can be done while the aircraft is slowing. We'll do the approach to the clean stall first. This will also give you an appreciation for the time it takes to decelerate a jet aircraft without the use of flaps. Set the power, or command the power to be set, at 50 percent N_1. Slowly increase angle of attack as the aircraft decelerates to maintain altitude. Notice that as the airspeed decelerates through V_{REF} + 20, the angle-of-attack indicator shows that we are at 1.3 times V_{S1}. It just so happens that 20 KIAS is the reference-speed adjustment applied when flying a flaps-inoperative approach: clever gadget that angle-of-attack indicator. As the needle enters the box, authoritatively apply power; in fact, shove the throttles all the way forward and then bring them back about an inch. Command the copilot monitor and adjust the power to preclude overspooling the engines. Simultaneously freeze the pitch attitude and command approach flaps. As the airspeed increases, modify the pitch attitude to maintain altitude. As the airspeed increases through V_{REF} + 10 KIAS, command flaps up, and at about 140 KIAS, reduce the power to at least 65 percent N_1 (Fig. 8-15).

1. **LEVEL FLIGHT**
   * Clean aircraft
   * Power - 50% N_1

2. **ENTRY**
   * Maintain altitude
   * Trim as required

3. **RECOVERY**
   * Power - Maximum allowable
   * Flaps - T.O. & APPR.
   * Pitch Attitude - Maintain
   * Speed - Increase to V_{REF} + 10
   * Flaps - Retract
   * Pitch - reduce to maintain altitude

NOTE:
Aircraft will have a nosedown pitch tendency due to engine spool-up. This may require additional back pressure on the yoke to maintain altitude.

Fig. 8-15. The approach to a clean-configuration stall and the recovery procedures in the Cessna Citation 500 aircraft require that the pitch attitude be frozen at the first indication of the impending stall.

There are three common mistakes made in the recovery from this approach to the clean stall. First, most pilots are not nearly aggressive enough with power application on the recoveries. The throttles should be moved briskly forward. It's the copilot's job to monitor the power setting and make sure that it does not exceed maximum allowable, and since your copilot for these maneuvers is likely to be an instructor pilot or pilot examiner, you can be fairly confident the job will get done. It is a proven fact that pilots under stress will revert to their earliest training. If you should ever get into an inadvertent stall in a jet, you will not want to be hesitant with power application.

Second, in this particular aircraft, as in several other jets with tail-mounted engines, power application, actually engine spoolup, causes a nosedown pitching moment. As we said, the pitch attitude should be frozen for the recovery. It will take attention to make sure that the nose does not drop when the engines spool up. This takes about 7 seconds to occur.

The third mistake pilots make in stall recoveries is letting the nose drop or actually pushing it down. Old habits die hard, and in every other aircraft most pilots have flown, stall recoveries involve reducing pitch attitude. As soon as the pilot realizes he or she has made this mistake, the pilot is likely to jerk the nose back up, and this almost invariably causes an accelerated stall as a secondary. Like everything else in jet flying, smoothness counts here.

The second approach to stall recovery will be done with the aircraft in the approach-flap configuration with the gear up. This recovery will also be from a 20° bank. The power will again be set at 50 percent N_1. As the airplane decelerates through 140 KIAS, begin a 20° banked turn in either direction. The airplane doesn't care. The absence of propeller slipstream effects makes the maneuver the same no matter which direction the turn is. As the needle enters the box, simultaneously roll the wings level, freeze the pitch attitude, and apply full power. Check that the flaps are at the approach setting. As the aircraft accelerates through V_{REF} + 10 KIAS, command flaps up (Fig. 8-16).

1. **LEVEL FLIGHT**
   * Flaps - T.O. & APPR.
   * Power - 50% N_1

   NOTE:
   Aircraft will have a nosedown pitch tendency due to engine spool-up. This may require additional back pressure on the yoke to maintain altitude.

2. **ENTRY**
   * Maintain altitude
   * Trim as required
   * As airspeed passes 140 KIAS roll smoothly into 20° bank

3. **RECOVERY**
   * Power - Maximum allowable
   * Flaps - Check at T.O. & APPR.
   * Pitch Attitude - Maintain
   * Wings - Level
   * Speed - Increase to V_{REF} + 10
   * Flaps - Retract
   * Pitch - reduce to maintain altitude

Fig. 8-16. The approach to an approach-configuration stall and the recovery procedures in the Citation 500 are similar to the clean-configuration recovery.

The same problems with recovery are usually noted with this recovery as with the recovery from the clean stall. Power application must be brisk. The manufacturer's acceleration tolerances for this engine state that if the throttles are moved from the idle position to the takeoff thrust position within 1 second, the engine can take up to 7 seconds to spool up to power. That can be a very long time with an airplane that is stalling. Any additional time taken to move the throttles will only add to this time period. Freeze the pitch attitude. This airplane will fly out of the maneuver with power application. As the airspeed increases, smoothly decrease the angle of attack to maintain altitude.

The last approach-to-stall recovery will be done in the full landing configuration. Perform the before-landing checklist and set the power to 65 percent N_1. Increase back pressure to maintain altitude. Notice that as the airspeed passes V_{REF}, the angle-of-attack indicator passes the 1.3 V_{S0} position. This is a good check of the accuracy of the V_{REF} calculation. After all, we know that the wing is going to stall at a specific angle of attack, no matter what the airspeed indicator says.

The recovery from this approach to a stall is a bit more complicated. First of all, as the needle enters the box, smoothly reduce the pitch attitude to the level-flight attitude as you apply maximum power. As soon as the airspeed indicator changes its direction of movement and the angle-of-attack gauge reverses its trend, command flaps to approach. At V_{REF} 10 KIAS, smoothly bring the nose up to the climb attitude, about 10° noseup, and begin a climb to regain any lost altitude. As soon as a positive rate of climb is noted, command gear up, and at V_{REF} +10 KIAS command flaps up. As soon as you have returned to the maneuver-entry altitude, smoothly modify the pitch attitude to maintain that altitude as the aircraft accelerates (Fig. 8-17).

This recovery seems easier for most pilots, probably because the nose is allowed to drop somewhat. The same admonitions apply here. Get the power up. Be smooth with the pitch. The nose should not be pushed down; the angle of attack just needs to be somewhat reduced. The aircraft is in a high-drag configuration, and power alone will not affect the recovery.

1. **LEVEL FLIGHT**
   * Gear - Down
   * Flaps - Land

2. **ENTRY**
   * Power - 65% N_1 (425 PPH)
   * Maintain Altitude
   * Trim as required

3. **RECOVERY**
   * Power - Maximum Allowable
   * Pitch Attitude - Reduce 2-3°
   * Wings - Level
   * Out of Buffet and IAS Increase - Flaps T.O. & APPR.
   * Airspeed Increase to V_{REF} - Rotate to 10° pitch
   * Positive Rate - Gear up
   * Climb - At V_{REF} to Entry Altitude
   * Accelerate to V_{REF} + 10 knots
   * Flaps - Up

NOTE:
Aircraft will have a nosedown pitch tendency due to engine spool-up. This may require additional back pressure on the yoke to maintain altitude.

Fig. 8-17. The approach to the landing-configuration stall and the recovery procedures in the Citation 500 require that the pitch attitude be slightly reduced at the first indication of the stall.

APPROACHES
Nonprecision approaches

It is important to note that the term nonprecision refers to the approach aid guidance to the runway, not the pilot's technique in flying these types of approaches. Speed control is vital on all approaches, but during nonprecision approaches, accurate and precise control of speed can make the difference between success and a missed approach. Never let anyone rush you when flying an airplane, and especially not when executing an approach.

The airplane should be slowed and configured so that by 1 mile outside the final-approach fix it is at the target speed for the approach (this will be no less than the minimum maneuvering speed for the approach configuration) with the gear down and the flaps at the approach setting. The speed will be the maneuvering speed for the approach-flap setting. You will need to convert this to ground speed in order to establish your timing for the missed approach point, if needed. The before-landing checklist should be complete except for final flaps and yaw damper. Establish the speed, trim the airplane, and determine the heading required to track the final-approach course PRIOR to the final-approach fix.

The secret to a successful nonprecision approach is to arrive at MDA well before the missed approach point with a trimmed and configured aircraft so that when the runway is in sight, it is a simple task to complete the landing checklist (landing flaps) and land the airplane. You will need to exercise good control of airspeed, attitude, and rate of descent. You should plan to be at the MDA when one half of the time computed to arrive at the missed approach point has passed.

Putting it all together, at the final-approach fix, note the time, reduce the power, and descend at your preselected and trimmed airspeed to the MDA. Instead of attempting to track the needles immediately upon crossing the final-approach fix, maintain your predetermined heading, which contains the proper wind correction, until you have descended to MDA and leveled the aircraft. Using this procedure will keep you well within course tolerances and in alignment with the runway 99 percent of the time. When 100–200 feet prior to MDA, smoothly bring the power back up to the setting that will maintain minimum maneuvering speed for the configuration. Because this is the speed at which you are descending and for which the aircraft is trimmed, the level-off at MDA should be accomplished with minimal trim change. With the runway in sight, make whatever small adjustments might be necessary, extend landing flaps, and descend for the runway. Again, to review:

1. Precompute the inbound dead-reckoning heading taking into consideration the known wind-drift correction.

2. Estimate ground speed, and determine time to missed-approach point and time to MDA.

3. Cross the final-approach fix at the proper altitude, at the predetermined airspeed with the aircraft configured and trimmed.

4. Descend to arrive at MDA well before the missed-approach point and at least no later than visibility limits.

5. Compute and fly a single heading for wind correction inside the final-approach fix.

As usual, the secret to success is staying ahead of the situation. Proper preplanning and organization in execution of nonprecision approaches is the key to consistently finding the landing runway and making a smooth transition to landing. Any final approach segment that seems frantic probably is, and the cause is probably lack of planning and organization. It has always amazed me that pilots can sit for hours and do absolutely nothing and then arrive at the final-approach segment of an instrument approach totally unprepared for the approach and landing.[8] Use that time to advantage! (See Fig. 8-18.)

The circling approach

Before we discuss circling approaches, I believe it is worth considering that most airline and many air-taxi-operator operational specifications prohibit the circling

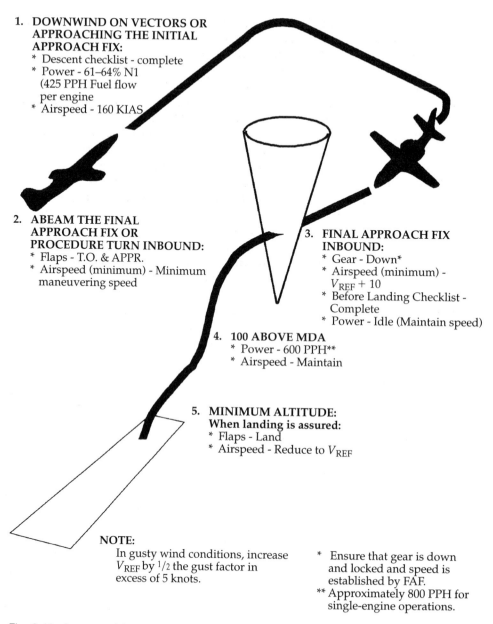

1. **DOWNWIND ON VECTORS OR APPROACHING THE INITIAL APPROACH FIX:**
   * Descent checklist - complete
   * Power - 61–64% N1 (425 PPH Fuel flow per engine
   * Airspeed - 160 KIAS

2. **ABEAM THE FINAL APPROACH FIX OR PROCEDURE TURN INBOUND:**
   * Flaps - T.O. & APPR.
   * Airspeed (minimum) - Minimum maneuvering speed

3. **FINAL APPROACH FIX INBOUND:**
   * Gear - Down*
   * Airspeed (minimum) - V_{REF} + 10
   * Before Landing Checklist - Complete
   * Power - Idle (Maintain speed)

4. **100 ABOVE MDA**
   * Power - 600 PPH**
   * Airspeed - Maintain

5. **MINIMUM ALTITUDE:**
   When landing is assured:
   * Flaps - Land
   * Airspeed - Reduce to V_{REF}

NOTE:
In gusty wind conditions, increase V_{REF} by $1/2$ the gust factor in excess of 5 knots.

* Ensure that gear is down and locked and speed is established by FAF.
** Approximately 800 PPH for single-engine operations.

Fig. 8-18. A nonprecision approach procedure requires that the aircraft be configured and trimmed at approach speed prior to the final-approach fix. Descent should be planned to arrive at MDA at least 1 mile prior to the missed-approach point.

approach unless the ceiling is at or above 1000 feet and the visibility 3 miles or greater. In other words, it must be basic VFR or a circling approach cannot be accepted. That should tell us something about what the professionals think of this maneuver.

Because the circling maneuver comes as the landing phase of a nonprecision approach, the aircraft should already be configured for the maneuver. The non-precision approach is conducted at the minimum maneuvering speed for the flap/slat configuration used. This speed will give adequate maneuvering airspeed for the circling maneuver. It should be noted that the visibility minimums and MDA selected for the circling approach should be based on this minimum maneuvering speed, NOT 1.3 V_{S0}. If your aircraft has a V_{REF} of 108 KIAS and the maneuvering speed with the approach-flap configuration is V_{REF} + 20 KIAS, your approach category will be Category C, not B as the V_{REF} would indicate. Circling minima are based on the airspace protected for each approach category.[9] See Fig. 8-19 for a representation of the airspace protected for the various approach categories.

The angle of bank should be such that the airport of intended landing is kept in sight at all times and the limits of the protected airspace are not transversed. In

Circling approach protected area

Approach category	Radius (NM)
A	1.3
B	1.5
C	1.7
D	2.3
E	4.5

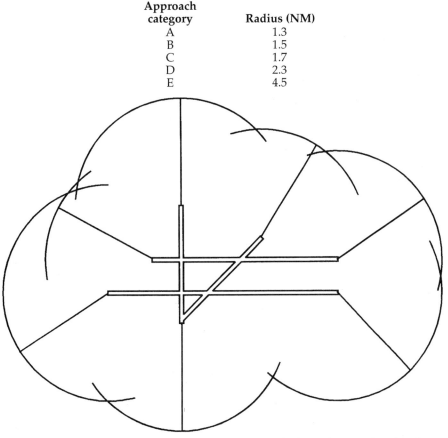

Fig. 8-19. The protected airspace for a circling approach depends upon the aircraft approach category. The airspace is defined by lines tangential to the airspace contained by the radius, defined by approach category, from the ends of all runways.

no case should more than a 30° angle of bank be used when maneuvering close to the ground. As a rule of thumb, use a 20° bank for speeds of 100 KIAS or below, 25° for speeds between 100 and 120 knots, and 30° for speeds in excess of 120 KIAS.

Maintain MDA exactly. It is not legal to be below this altitude until in the position to make a normal descent for landing. Drifting above this altitude is not only sloppy flying, it might put you back into the clouds. Extend your downwind 15 seconds past the threshold of the landing runway. Make a continuous turn from downwind to final. A 3-degree descent will put you 300 feet above the threshold for every mile from the airport you are. Landing flaps should be extended when the landing is assured. When the aircraft is established on final approach with landing flaps, the speed should be reduced to V_{REF}.

The circling maneuver requires extreme vigilance on the part of both crewmembers. This maneuver, which requires visual reference at all times, is only a quasivisual maneuver. In most cases, there will be a low ceiling and a restriction to visibility. The pilot will be flying partially by reference to instruments and partially by ground reference. The potential for spatial disorientation is high during this type maneuver. The nonflying pilot must closely monitor altitude, airspeed, angle of bank, and rate of descent. Be careful that an excessive sink rate does not develop and that the aircraft does not end up on the adverse side of the power curve when extending landing flaps.

The ILS approach

A successful ILS approach, just as a successful nonprecision approach, is the result of careful preplanning. An ILS approach, however, is somewhat easier since it requires fewer power manipulations and is, of its nature, a steady and stabilized maneuver. Just as for a nonprecision approach, the pilot should have a wind correction heading already nailed down prior to passing the final approach fix inbound. The proper descent rate for the localizer is easily computed, and it is printed on Jeppesen approach charts.

When being vectored for an ILS approach, strive to arrive 3–4 miles from the final-approach fix with the flaps set for approach and the speed stabilized on and trimmed for minimum maneuvering speed with approach flaps. You should also, if possible at this time, be at the glide-slope intercept altitude. (If you are flying a full approach with a procedure turn, this would be the procedure turn inbound.) The landing checklist, with the exception of gear and final flaps, should be complete at this time. At one dot from glide-slope intercept (below), extend the landing gear and select landing flaps when crossing the final-approach fix. At this point, establish the proper rate of descent for the glide path, and adjust power to maintain V_{REF} + 10.

All that sounds easy doesn't it? But what about the part regarding the proper rate of descent for the glide slope. Well, that's a figure you can read from the approach chart, but you can also compute it, and knowing how to do this is more useful than you might believe. A 3° glide slope is about 300 feet per nautical mile of descent. If you know your ground speed (and you should!), you can compute the required rate of descent. Just take your ground speed, divide it by two, and add a zero. That's the number of feet per minute descent you need to maintain to make good the glide slope. It looks like this:

Ground speed = 120 knots
Divided by 2 = 60
Add a zero = 600 fpm

This is also very useful in the reverse. Let's suppose you are flying an ILS approach with an indicated airspeed of 120 KIAS. You are maintaining the glide slope with a 300-fpm rate of descent. What does that tell you? Well, work the problem in reverse.

Feet per minute = 300 fpm
Take off a zero = 30
Multiply by 2 = 60

The ground speed is 60 knots. So what's the reason for knowing that? Well, if your ground speed is only 60 knots, and your indicated is 120, that means you are flying an ILS into a 60-knot headwind. If anything at all happens to that headwind, you are going to be in supreme difficulty.[10]

One of the major reasons for insisting on a stabilized approach is so that it becomes very apparent when the approach becomes unstabilized. An unstabilized approach inside the outer marker is ample reason to initiate a missed approach. If, however, the airspeed and rate of descent are fluctuating all over the gauges because you have not stabilized the approach, how will you know if the instability is a result of your errors or a serious windshear?

A target rate of descent for a glide slope is 500 to 600 feet per minute. That takes into account ground speeds from 100 to 120 knots. You can compute the rate for any speeds outside that range. Initiate the computed rate of descent, and the minute the glide-slope needle moves away from the center, modify the rate. It shouldn't take much if you catch it immediately. You should already have turned to the heading you know will nail the localizer, so the approach should now just be a matter of SMALL corrections to allow for changing winds as you descend.[11]

The quickest way to destabilize an approach is to be constantly making power adjustments. You should have an idea of the power settings required to achieve a specified airspeed in a specified configuration. Bracket your power changes around this target setting and give the engines a chance to respond. Getting anxious and not waiting for the airplane to respond will almost guarantee that you will quickly be 180 degrees out of phase with the power requirements. Because changes in power can also cause pitch changes in some aircraft, maintaining a constant rate of descent to track a glide slope might become pretty well impossible. A two-engine stabilized ILS approach should bring you to minimums on the glide slope at V_{REF} with the aircraft configured for landing. All that is necessary is to fly to the runway, reduce power as recommended by the manufacturer, and stop the aircraft. Good thing, too, because at 200 feet and ½ mile in actual low IFR conditions, there's not much time to do much else other than execute a missed approach.

You might have noticed that nonprecision approaches are conducted using only approach flaps until the landing is assured, but normal ILS approaches are conducted with landing flaps from the marker inbound. That is done for good reason. The power required to arrest a maximum-rate descent from the final approach fix to MDA with full flaps is excessive. Maneuvering with full flaps will certainly keep the engines spooled up, but it will also waste fuel, upset the passengers, and set off every noise sensor in the county. The stabilized ILS approach with landing flaps is designed to put the aircraft in the best configuration for landing from a low-visibility approach with minimum effort. There is no intermediate level-off as with a nonprecision approach, and the maneuvering phase of the approach is over by the time landing flaps are extended (Fig. 8-20).

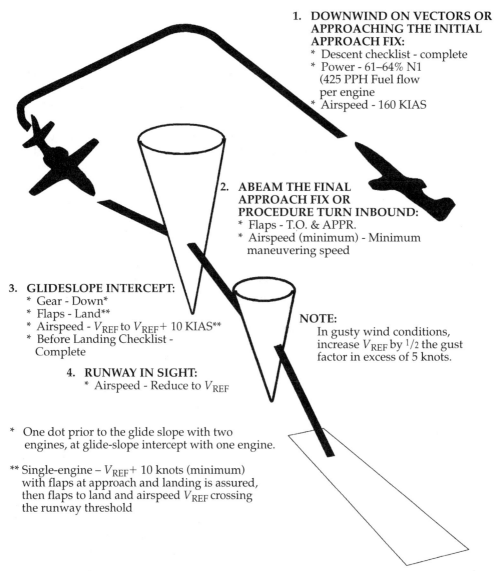

1. DOWNWIND ON VECTORS OR APPROACHING THE INITIAL APPROACH FIX:
* Descent checklist - complete
* Power - 61–64% N1 (425 PPH Fuel flow per engine
* Airspeed - 160 KIAS

2. ABEAM THE FINAL APPROACH FIX OR PROCEDURE TURN INBOUND:
* Flaps - T.O. & APPR.
* Airspeed (minimum) - Minimum maneuvering speed

3. GLIDESLOPE INTERCEPT:
* Gear - Down*
* Flaps - Land**
* Airspeed - V_{REF} to V_{REF}+ 10 KIAS**
* Before Landing Checklist - Complete

NOTE:
In gusty wind conditions, increase V_{REF} by 1/2 the gust factor in excess of 5 knots.

4. RUNWAY IN SIGHT:
* Airspeed - Reduce to V_{REF}

* One dot prior to the glide slope with two engines, at glide-slope intercept with one engine.

** Single-engine – V_{REF}+ 10 knots (minimum) with flaps at approach and landing is assured, then flaps to land and airspeed V_{REF} crossing the runway threshold

Fig. 8-20. An ILS approach should be a stabilized maneuver. The only job that the pilot should have from the final-approach fix inbound should be flying the needles. All checklists and configurations should have been accomplished prior to this point.

Engine-out approaches

Engine-out nonprecision approaches are executed just like all-engine nonprecision approaches; the speeds, configuration, and procedures are the same. The only difference will be the power selected on the remaining engine.

Engine-out ILS approaches are conducted differently from the all-engine variety, however. That difference is a difference in flap configuration only. Airspeeds, gear extension, and rate of descent are standard; only the power setting on the remaining engine is nonstandard. The flaps configuration on an engine-out ILS approach specifies approach flaps until the landing is assured, and then landing flaps are selected. Selecting landing flaps at the marker as is done with all engines operating would make a missed approach much more difficult. The power required to transition from a descent to a climb in this configuration might be more than what is available on a single engine.

Missed approach/rejected landing

A pilot's mind is always set on completing an instrument approach and landing the airplane if the weather is reported to be at landing minimums or better. A missed approach or a rejected landing is always a possibility, however. A good pilot will always be prepared for that eventuality, whether it be due to weather, an unsafe gear indication, or the traditional "truck on the runway." Missed approaches and rejected landings are required on the rating checkride.

Although there have been several fatal accidents during the rejected-landing/missed-approach phase of flight, I have always considered this a safe maneuver if executed properly. How is it possible to get hurt if you do not descend below MDA or DH? The secret, of course, is to be mentally prepared for the missed approach and then to make the timely decision at MDA and the missed-approach point or DH and stick by that decision.

Prior to beginning an instrument approach, the pilot should always review the approach plate and be familiar with the missed-approach procedure. No one is expected to memorize the total missed approach procedure, and with a crew of two this certainly is not necessary. The most important information, how high and which direction, should be committed to memory. It is necessary to know the direction of the required initial climb since this direction is usually dictated by obstacle clearance considerations. The nonflying pilot can set up the proper headings and radials after that point.

A missed approach or a rejected landing is nothing but a takeoff started a little higher and a little faster than normal. The 1.3 V_{S0} used as V_{REF} for the landing approach in a jet is usually close enough to V_2 to make the difference not worth noting. The missed approach might not be from the landing-flap configuration, but the rejected landing certainly will be. The procedure will be basically the same for both maneuvers, however. As soon as the decision is made to miss, the power should be authoritatively advanced to the takeoff/go-around setting as the aircraft is rotated to the climb attitude. If you are flying with a flight director, hit the go-around switch, and rotate according to the flight director's commands. You can modify the pitch attitude as soon as you have the climb established; it is preset to be safe for single-engine climb and will probably be too shallow on two engines. The pilot flying should command flaps to the takeoff position, and when a positive rate of climb is achieved, the gear should be retracted. The aircraft is in the same configuration as every other takeoff, and remaining flaps should be retracted according to the normal flap-retraction schedule (Fig. 8-21).

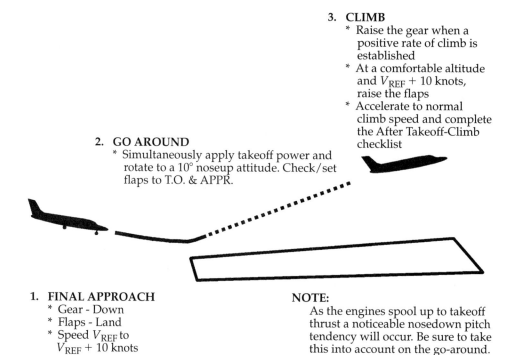

3. CLIMB
* Raise the gear when a positive rate of climb is established
* At a comfortable altitude and V_{REF} + 10 knots, raise the flaps
* Accelerate to normal climb speed and complete the After Takeoff-Climb checklist

2. GO AROUND
* Simultaneously apply takeoff power and rotate to a 10° noseup attitude. Check/set flaps to T.O. & APPR.

1. FINAL APPROACH
* Gear - Down
* Flaps - Land
* Speed V_{REF} to V_{REF} + 10 knots

NOTE:
As the engines spool up to takeoff thrust a noticeable nosedown pitch tendency will occur. Be sure to take this into account on the go-around.

Fig. 8-21. A missed approach must be a positive maneuver. It is necessary to establish and maintain a positive rate of climb throughout the maneuver.

This maneuver has some common mistakes associated with it. The two most serious that I see are not advancing the power fast enough and not climbing. You must be forceful in initiating the missed approach. The aircraft might be as close as 200 feet AGL when the decision to miss is made. It will be somewhat lower than that when the message gets to the hands and is executed. Get the engines making power; remember that there will be a lag even after you advance the throttles. Get the aircraft climbing; remember that as the engines develop power, a nosedown pitching moment results in many aircraft. This is going to counter your efforts to climb, and you must be ready for it. Also, if an immediate turn is required, start that turn as soon as speed permits.

A single-engine missed approach will be conducted in much the same manner except that it will of course have the added attraction of directional-control concerns. Keep the wings level and keep the ball in the center and climb. Of course, the flight director in the go-around mode will command the proper pitch attitude. Flight director go-around modes will command a pitchup consistent with V_2 on one engine. Once the gear is up and the flaps are at the takeoff setting, the aircraft is in the same configuration as on a takeoff with an engine failure and the same procedures will be used (Fig. 8-22).

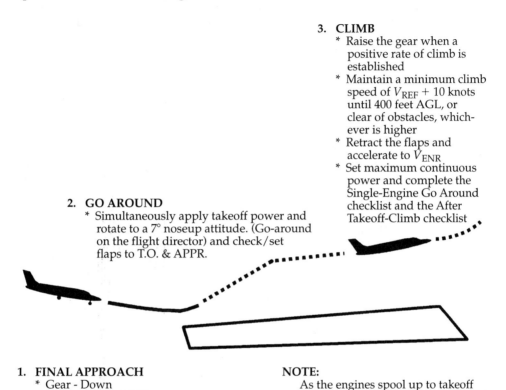

3. CLIMB
* Raise the gear when a positive rate of climb is established
* Maintain a minimum climb speed of V_{REF} + 10 knots until 400 feet AGL, or clear of obstacles, whichever is higher
* Retract the flaps and accelerate to V_{ENR}
* Set maximum continuous power and complete the Single-Engine Go Around checklist and the After Takeoff-Climb checklist

2. GO AROUND
* Simultaneously apply takeoff power and rotate to a 7° noseup attitude. (Go-around on the flight director) and check/set flaps to T.O. & APPR.

1. FINAL APPROACH
* Gear - Down
* Flaps - T.O. & APPR.
* Airspeed (Minimum) - V_{REF} + 10 knots

NOTE:
As the engines spool up to takeoff power a noticeable nosedown pitch tendency will occur. Be sure to take this into account on the go-around.

Fig. 8-22. A single-engine missed approach must also be a positive maneuver. A positive rate of climb must be maintained throughout the procedure.

LANDINGS

Landings, unlike takeoffs, are given considerable attention during flight training, so it is likely that you will, by this time, have received considerable instruction and practice in landing techniques. Okay, well enough! A jet is just another airplane. The primary difference between the landing for the jet and that for a similar-sized propeller aircraft is that the jet pattern should be about 50 percent larger to allow time for deceleration. (Remember, you no longer have those big rotating speed brakes to help you.) Some things will never change, however. A good landing is the result of a good approach, and a good approach is the result of proper planning.

Enter the landing pattern downwind in the clean configuration. The descent and prelanding checklists, with the exception of gear and flaps, should be completed at this point. Establish the initial approach power for your aircraft. We'll use the Citation figures again since this is a common entry-level jet, so set 63–65 percent N_1 (or 400–425 pph fuel flow, if you prefer). By the way, this power setting will stabilize the aircraft at 160 KIAS with a clean wing at approach altitudes. When opposite the midpoint in the runway, extend approach flaps, and be ready for the resultant pitch change. Your minimum speed in this configuration is V_{REF} + 20 KIAS.

Opposite the point of intended landing, command gear down. Proceed downwind to the 45° point past the end of the runway. At this point, initiate a descending turn to base, adjusting power to maintain V_{REF} + 20 KIAS. This should be a minor adjustment. When turning final, command landing flaps and adjust the speed to V_{REF} + 10 KIAS. The rate of descent should be consistent with a 2.5° to 3° descent in the neighborhood of 500–600 fpm. When landing flaps are extended, the power will need to be advanced to 63 percent N_1 to maintain speed and rate of descent.

As you approach the end of the runway, begin a slow deceleration to V_{REF}. The end of the runway should be crossed at 50 feet with the landing target 1000 feet down the runway. DO NOT DUCK UNDER. Flight instructors start hammering away at pilots from their first hours of landing practice that a good landing is "on the numbers." Well, that's just not true in the jet. To put the airplane on the numbers from a stabilized approach, you have to destabilize it! A jet is landed "by the numbers" not "on the numbers." The landing point is 1000 feet down the runway, and the threshold is crossed at 50 feet in a 3° descent. Period. Landing-performance charts for jets are based on the approach being flown this way. You'll surely fail a checkride, at the least, if you persist in the bad habit of ducking under. If you start flying larger airplanes, you might find yourself dragging gear in the approach lights.

After crossing the threshold, reduce power according to the manufacturer's recommendations. In the Citation, that means power to idle at 50 feet. If you don't, it just won't land. Hold the aircraft in the landing attitude and touchdown. Follow the recommended post-touchdown procedures. The usual sequence is power to idle, speed brakes/ground spoilers extend, maximum brakes, and thrust reversers, if installed. In the Citation aircraft, it is vital to have the nosewheel firmly planted on the pavement before deploying thrust reversers. The noseup pitching moment of thrust reverser deployment can lift the nosewheel off the pavement if it is not firmly held there.

Recall that the most common mistakes made during landings are undershooting or ducking under on final and floating. We've already discussed ducking under, but the float needs some consideration. Speed control is extremely important in the approach, flare, and touchdown.

If the airspeed is inadvertently high and a go-around is not initiated, it is still extremely important that the touchdown be made as close as possible to the 1000-foot point. You might have to make a conscious effort to put the aircraft on the ground and land a little faster and harder than normal. The natural tendency is to hold the aircraft off the runway, bleed speed off, and wait for a nice soft landing.

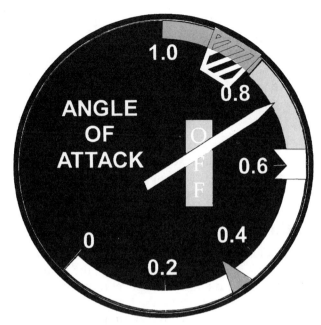

Fig. 8-23. This angle-of-attack indicator is indexed so that 1.0 corresponds to the stalling angle of attack. An indication of 0.6 will correspond with 1.3 times the stalling speed in the existing wing configuration.

This is an excellent way to become acquainted with the grass off the other end of the runway.

A jet is inherently aerodynamically clean, and even with full landing flaps, a jet will not experience as much drag as will a propeller aircraft in the landing phase. I'll say it again, you'll just never appreciate how much drag those props are until they are not there. A jet aircraft in ground effect can float for a considerable distance and might end up using more runway than is comfortable or available for stopping. Put the airplane down as near to the 1000-foot target point as is possible and use brakes, speed brakes, and thrust reversers to decelerate. You will be safer every time if you follow this procedure.

If you are too high crossing the end of the runway rather than too low or on target, you will of course need additional runway to bring the aircraft to a stop. If you are on speed and cross the runway threshold at 100 feet AGL rather than at 50 feet AGL with a normal 3° glidepath, you will use an additional 900 feet before touchdown, which is 900 feet beyond the 1000-foot target aiming point.

Once the aircraft is on the ground, it must be stopped. A factor that greatly affects stopping distance is the coefficient of friction between the tires and the runway. Many factors can have an effect on this, but the main ones are obviously runway condition and tire condition. Damp, wet, slush-covered, and icy runways will, of course, increase landing distance. Tire inflation is important on contaminated runways since the hydroplaning speed will be nine times the square root of the tire pressure. Any speed above this speed will present the possibility of hydroplaning. A tire inflated to 100 PSI will hydroplane at 90 knots. A tire underinflated at 80 PSI will hydroplane at 80 knots.

The aircraft weight affects stopping distance. The kinetic energy, or the energy of motion, possessed by a moving object is one-half the mass multiplied by the velocity squared:

$$KE = \frac{1}{2} MV^2$$

If the mass is increased, the kinetic energy is increased. The energy required to stop the object is equal to the kinetic energy. Notice, also, that the energy required to stop will vary as the square of the speed. If you double the speed, you will quadruple the energy required to stop.[12]

Rules of thumb for landings

As you know from looking at the landing-field length charts, they give the total distance from the 50-foot point above the threshold to the point where the aircraft is stopped on the runway. This is all for a dry, level runway with the aircraft being at the correct speed and altitudes. Because the target aiming point is 1000 feet down the runway, you know right away that the ground roll is going to be the total chart distance minus 1000 feet. There are some variables we can take into account to calculate some realistic stopping distances for the aircraft. After the 1000 feet has been deducted, apply the following to arrive at a more realistic rollout distance:

- For each knot above V_{REF}: Add 2 percent to rollout distance
- For 100 feet instead of 50 feet at threshold: Add 900 feet
- For each 10 knots headwind: Decrease 15 percent
- For each 15°F change from ISA: 4-percent change in landing roll
- 1° uphill runway slope: 4-percent decrease in landing roll
- 1° downhill runway slope: 6-percent increase in landing roll

These rules of thumb will give you the landing roll corrected for variations in technique, wind, temperature, and runway slope. Now let's see how this needs to be adjusted for runways other than dry. We will consider a wet runway to be any concrete runway that is wet and that has a braking action reported as good. A wet and slippery runway is any asphalt or blacktop runway that is wet, or any runway that is heavily trafficked and has rubber deposits, and the like, mixed with water on the runway, or any snow-covered runway, and finally any surfaced runway with the braking action reported as fair. An icy runway is any runway covered with ice or for which the braking action has been reported as poor to nil.

The FAA lumps all these runway conditions together and applies a runway increase of 15 percent to allow for all these conditions. Believing that can put you in the weeds off the end. The following correction factors are more realistic:

- Wet: Add 25 percent to the ground roll.
- Wet and slippery: Add 60 percent to the ground roll.
- Icy: Add 125 percent to the ground roll.

Let's add it all up. You are flying your favorite jet into an airport that has heavy jet traffic. Rain has just stopped, and you plan to land on a 5700-foot runway. You get a slam-dunk visual approach and are 50 feet high and about 5 knots fast over the threshold. The landing-distance charts say you will need 3000 feet for this landing. You have 2700 feet to spare, right? Wrong! It adds up like this:

- Runway required per charts   3000
- Less 1000 aiming point   −1000
- Rollout distance   2000
- 10 knots excess airspeed = +20 percent (feet)   +400
- Total rollout   2400
- Multiplication factor for wet and slippery ×   +0.60
- Addition for wet and slippery   +1440
- Total rollout distance required   3840
- Total Runway available   5700
- Distance to touchdown point   −1000
- Distance to touchdown due to being high   −900
- Total available   3800

It appears that in this instance you will be explaining to the FAA and the insurance company just how you managed to run off the end of the runway when the landing distance charts said you had more than enough runway to land and stop the airplane. None of these factors in and of themselves seem to be a major problem, but like much else in aviation, the effect quickly snowballs, and you find yourself in a troubling situation you didn't anticipate. Fly by the numbers.

If you use normal braking rather than maximum braking energy, you can expect to use about 60 percent more runway than charted for the rollout. If your antiskid is inoperative, add 75 percent. Neglecting to use ground spoilers will cost about 25 percent.

ANGLE OF ATTACK

We have talked about angle of attack and the aerodynamics of the wing and have touched on the angle-of-attack indicators installed on the HSI and ADI. This is one of the most useful gauges in the cockpit of a modern jet aircraft. It can be calibrated in many ways with numbers, indices, or simply different colors. No matter what the design, however, each tells the same story: how close the aircraft is to the stalling angle of attack. This is a good crosscheck for your calculation of V_{REF}. No matter what the airspeed indicator says, we know that a wing will stall at a predetermined stalling angle of attack. This angle is set when the wing is designed, and ambient conditions do not vary that angle.

The indexed point on the angle-of-attack indicator tied to 1.3 V_{S0} is also useful during takeoff. Because V_2 is approximately 1.25 V_{S1}, the angle-of-attack indicator is also a good crosscheck for V_2. Angle-of-attack gauges, if properly indexed can also be used for maximum-range cruise since this is also a function of angle of attack. A representative angle-of-attack indicator is shown in Fig. 8-23. The index at the 0.6 point (3 o'clock) on the indicator correlates with V_{REF}. The index at the 0.35 (4:30 position) indicates the angle of attack that will result in the most miles per gallon, which is the maximum-range speed.

No matter what version of angle-of-attack indicator you have in your aircraft nor what displays are in the cockpit, get this instrument incorporated into your scan, especially on landing approach. It will tell you the true story.

Endnotes

1. A transport-category aircraft is simply any aircraft certified under FAR 25. Most small general aviation aircraft are certified under FAR Part 23.

2. A transport-category aircraft flight manual is different from the pilot operating handbooks found in FAR 23 aircraft. The FAR 25 flight manual contains normal, abnormal, and emergency procedures, plus performance charts for takeoff, initial climb, and landing. Cruise performance charts and systems descriptions are not found in these manuals. The charts and descriptions are included in a separate book that is not part of the required aircraft documentation.

3. This is true for all two-engine turbojets, be they Citations or 757s. Three- and four-engine turbojets must, understandably, demonstrate better performance; because most entry-level jets have two engines, we will address only that performance. Refer to FAR 25 for more information.

4. There is nothing special about jet weight and balance. If you can compute center of gravity for a Cessna 172 or Cherokee Arrow, you can do the calculation for a jet. The numbers are just larger, not more complex.

5. Try stopping a piston twin on the runway someday with the engines set at the zero thrust setting rather than at idle if you want to begin to appreciate this effect. My flight instructor, although a very experienced pilot at the time, gained full appreciation of the decelerating effect (or lack of) of idling props when he landed a Twin Comanche with both engines feathered. He had the engines feathered and the props horizontal because he was landing with a gear malfunction and was trying to save the engines. He remarked later that he was not prepared for just how far that airplane floated without the drag of the props.

6. The location of the fuel dumping area and the time that is necessary to dump the fuel required to reduce the gross weight to maximum landing weight would have been determined during flight planning, of course.

7. Every aircraft has a minimum maneuvering-speed schedule that is dependent upon flap/slat deflection. This is the minimum speed to be used when maneuvering during takeoff or landing phases of flight. When the aircraft is established on final, this speed is abandoned in favor of V_{REF}. Minimum maneuvering speeds will usually be expressed in relation to V_{REF} since the relationship to *stalling speed* is of interest to the pilot, not the exact number on the airspeed dial.

8. Perhaps this comes from the fact that the FAA seems to want to insist that everything that happens on a checkride come as a total surprise. When pilots have this reinforced through many checkrides, they might come to believe that the approach should be a surprise procedure to be conquered against overwhelming

odds. If that is so, perhaps the FAA needs to change its philosophy.

9. Circling visibility minimums and protected airspace increase with increasing approach speed because the turning radius of an airplane is based on the square of the airspeed divided by 11.26 times the tangent of the bank angle.

10. This happened to a hapless airline crew. The 60-knot headwind abruptly changed to an approximate 10-knot tailwind at 200 feet AGL. They were at 200 feet and ½ mile from the runway on an ILS approach and didn't get there! Not much was known about microbursts at the time this accident happened.

11. Just as a rule of thumb, the winds 1000 to 2000 feet above the runway are usually about 90° to the left of the surface winds and about twice as strong. If you know the surface winds, and you should have that information, you will be prepared for the wind at the outer marker and the corrections that might be likely while executing the approach.

12. This relationship of stopping energy to speed should be remembered. When you have plenty of runway available, it is more cost efficient to delay heavy braking until the aircraft has slowed somewhat. If you have reversers and can use them to decelerate from 120 to 60 knots before you apply the brakes, you will incur only ¼ the brake wear!

9

Cockpit resource management

WHY THERE ARE TWO

THERE IS ONE VERY GOOD REASON FOR HAVING TWO PILOTS IN THE cockpit of a modern jet, and it is the best of all reasons: *They are both necessary.* Because both are necessary (and not just because of aircraft certification or 121 regulations), both are equally important in the whole flying equation. The days of the autocratic captain barking orders and expecting that every wish and whim be obeyed and satisfied without question are (hopefully) gone.

The copilot is not just an apprentice, but a member of a team whose goal is the safe and efficient completion of the flight assignment. The seat occupied does not signify intelligence or value—these days it might not even be a valid indicator of relative experience—but is only a reliable indicator of company seniority. Any well-functioning team must have a captain. There must be one who has the final authority and responsibility for the flight. The regulations require this and common sense dictates it.

In most training programs, both 121 and corporate, the copilot undergoes the same training as the captain. The only difference being the checkride taken at the end of training. The captain completes a pilot-in-command check and the copilot completes a second-in-command check. In airline flying, the copilot is not usually type rated in the aircraft until his or her seniority number comes up and he or she is able to upgrade to captain, but the copilot is nonetheless expected to be just as knowledgeable about aircraft systems and just as proficient in flight as the captain. In corporate flying, this is being replaced by a trend toward having cocaptains in the aircraft with each pilot being a type rated captain.

No matter which system is being followed, the folks in the front seats should be considered and should consider themselves as members of a team whose united goal is the safe and efficient completion of the flight assignment.

INFORMATION SHARING

One of the key elements in the functioning of an efficient and coordinated team is the sharing of all pertinent information. There should never be any question in either pilot's mind as to the mission to be completed or the means to be used to accomplish that completion. The progress of the mission should be agreed upon by both pilots, and either should be able to take over for the other at a moment's notice. Not only should each have and be aware of specific responsibilities, but each should be aware of the responsibilities and progress of the other.

This all sounds really good in theory, but how in the world is this put into effect and practiced in the aircraft cockpit? Several practices and procedures enhance and promote crew coordination and cockpit resource management. One of the most important practices is the standardization of training and procedures.

Standardization

It is always satisfying to see two pilots from opposite sides of the country get into the simulator and function as a smoothly coordinated crew on the first simulator session. This can happen only through standardized training. Both were trained in the same manner and to the same standards and each knew what to expect of the other during any phase of flight.

One area where standardization is vital is callouts required during flight. If both pilots agree on what should be called and when, there will be a system of checks and balances. If there is no agreement, there will be confusion since one pilot will make an unanticipated call or misunderstood call, or one pilot will omit a call the other thought was vital. The front end of a jet aircraft boring through the fog at well over 100 knots toward a hard slab of concrete is no place for confusion.

Table 9-1 (on pp. 266 and 267) shows a list of recommended callouts for various phases of flight. This is by no means presented as the only authoritative list, and I am sure it will generate much discussion. At any rate, all callouts should be acknowledged by the other pilot. Lack of acknowledgment should indicate lack of receipt of the communication, and the callout should be repeated. Proper acknowledgment does not always have to be verbal. The callout "V_1" can be considered acknowledged when the pilot places both hands on the yoke. An acceptable acknowledgment to V_R would be rotation of the aircraft to the takeoff attitude. Callouts of deviations from course, heading, speed, and altitude, however, should immediately be acknowledged by the other pilot. It should be agreed between the pilots what the acknowledgment should be, but the simple word "correcting" accompanied by the appropriate action would seem to be adequate.

Crew briefings

A before-takeoff briefing should be accomplished before receiving a takeoff clearance. This briefing should contain a narrative of just how the takeoff and initial climb will be conducted and should set out the responsibilities of each pilot. The following briefing is one I have used that has always worked well. Again, this is open for modification as long as the substantially same information is included.

This will be a (static/rolling) takeoff with flaps set to (takeoff and approach/zero).

I will advance the throttles, you will set takeoff power to (precomputed takeoff power setting.)

Call airspeed alive, 80 knots, V_1, rotate, V_2, and positive rate. I will command gear up, and you will raise the gear. Call $V_2 + 10$, and I will command flaps up.

Monitor all instruments and the annunciator panel during takeoff. At the 80-knots call, crosscheck both airspeed indicators. In case of serious malfunction before V_1, call abort in a clear, loud voice. (Captain can reserve the right to call abort.)

If a malfunction occurs at or after V_1, we will continue the takeoff. Advise me of the malfunction, and we will handle it as an inflight emergency. Plan to fly (state intentions for return to airport).

Departure instructions for this takeoff are: (Initial departure instructions or SID)

Radios are set to (state frequencies, etc.).

Any questions?

Although the above briefing might seem to be a lot to remember, just think of the events as they occur in the course of a normal takeoff. If it is certain that both pilots understand the items in a standard briefing, items that are prior to the plan of action for an immediate return may be covered with the phrase "standard briefing."

Briefings are not limited to takeoff briefings, however. Approaches should be briefed just as thoroughly to preclude any misunderstanding and to catch any errors either pilot might have made in receiving the clearance or setting the radios. The approach briefing, which should be accomplished prior to the before-landing checklist—if possible, prior to approach vectors or arrival at the initial-approach fix—should accomplish the following:

1. Identify the approach to be flown and the transition.
2. Assign the nonflying pilot to identify all intersections and the final-approach fix.
3. Assign the NAV frequency and course selection changes for the entire approach and initial missed-approach segment.
4. Assign the timing responsibility, if appropriate.
5. Assign the standard callouts:
   a. 1,000 feet before assigned altitudes
   b. Localizer or course alive
   c. 500 feet and 100 feet above published minimums
   d. Minimums, runway *not* in sight, or
   e. Minimums, runway in sight, cleared to land.

Table 9-1. Standard crew callouts

LOCATION	CONDITION	CALLOUT
TAKEOFF	Computed N_1 Set	"Power set"
	Both airspeed indicators move off the peg	"Airspeed alive"
	Both airspeed indicators indicating 70 KIAS	"70 knots"
	Airspeed indicators at computed V_1	"V_1"
	Airspeed indicators at computed V_R	"Rotate"
	Airspeed indicators at computed V_2	"V_2"
	Positive rate of climb indicated	"Positive rate"
DEPARTURE/ ENROUTE/ APPROACH	Prior to intercepting an assigned course	"Course alive"
	1000 feet prior to level off	State altitude leaving and assigned level-off altitude
CLIMB AND DESCENT	Approaching transition altitude (IFR)	"Transition altitude, altimeters set"
	1000 feet above/below assigned altitude (IFR)	"State altitude leaving and assigned level-off altitude"
FINAL (IFR)	At final approach fix	"(FIX) altimeters and instruments check."[1]
	500 feet above minimums	"500 feet above minimums"
	100 feet above minimums	"100 feet above minimums"
	Visual reference required by FAR 91.175(c) is continuously established[2]	"Runway at (clock position" or "Approach lights at (clock position)"
	After pilot flying reports "Visual" pilot not flying reverts to instruments and callouts	"V_{REF} + airspeeds" "Sink (rate of descent)" "On," "Above," or "Below glide slope," if appropriate
	At DH (decision height)	"Minimums, runway not in sight," or "Minimums"

Table 9-1.

LOCATION	CONDITION	CALLOUT
		runway at (clock position)" or "Minimums, approach lights at (clock position)"
	At MDA	"Minimums"
	AT MAP	"Missed approach point, runway not in sight," or "Missed approach point, runway at (clock position)", or "Missed approach point, approach lights at (clock position)"
FINAL (VFR)[3]	500 feet above field elevation	"500 feet above field"
	100 feet above field elevation	"100 feet above field"

1. Check for appearance of warning flags and gross instrument discrepancies. Captain's judgement on excessive altimeter error.
2. Care must be exercised to preclude unnecessary or premature callouts that can adversely influence the pilot flying and result in a premature abandonment of instrument procedures.
3. It is recommended that all crews use electronic and/or visual systems as an aid in maintaining position on the glide slope.

The following sample approach briefing is based on the ILS Rwy 16R approach to Van Nuys, California. See Fig. 9-1 for a sample of the chart for this approach.

This will be an ILS to runway 16R at Van Nuys. Please tune and identify the Van Nuys localizer, 111.3 in NAV 1 and set 161° in the course selector window. Also tune and identify Fillmore VOR, 112.5 on NAV 2 and select 053° to identify UMBER. After we have intercepted the localizer, please tune and identify the Van Nuys VOR, 113.1 on NAV 2 and set the DME to NAV 2 to identify KADIE. After the DME locks onto the VNY VOR, set the DME to HOLD and tune NAV 2 to the ILS frequency and make your HSI match mine. In case of a missed approach, I will climb straight ahead to cross the VNY 1.5 DME at or below 1750, then a climbing left turn to 4000 feet to intercept the VNY R-101. DH for this approach is 1040 barometric, 250 on the radar altimeter. V_{REF} for this approach will be 103 knots and the go-around power will be 93.8 N_1. Please give me all standard callouts. Any questions?"

It should be noted that the pilot giving this briefing will not be in command of the airplane at the time. Contrary to the FAA's position for initial instrument ratings, it is not possible to fly an airplane and read an approach plate at the same time. The captain will relinquish physical control of the aircraft to the copilot before conducting the briefing. After the briefing is complete, the captain will resume physical control, and the copilot will set up the radios as briefed.

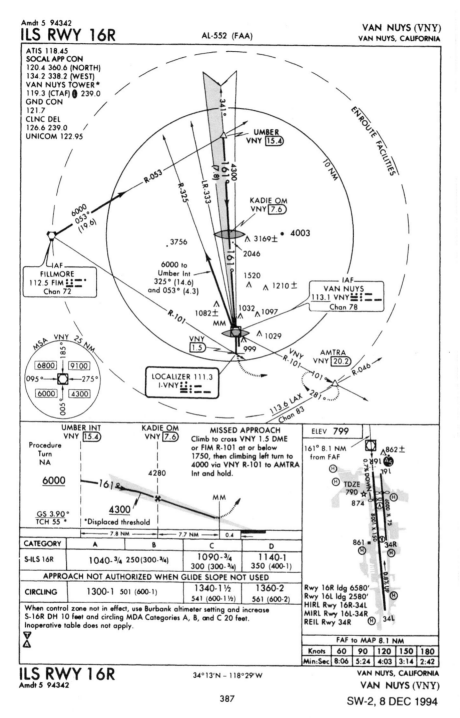

Fig. 9-1. The standard instrument approach procedure chart should be referred to for the approach briefing. This is the Runway 16R approach to Van Nuys, California (VNY).

268   Cockpit resource management

One pilot must always be in physical control of the aircraft, and transfer of this control should be a definite event. One pilot must acknowledge relinquishing control while the other pilot must acknowledge taking control. There has been more than one incident after which the pilots looked at each other and said in chorus "I thought you were flying!"

These are the standard briefings that should be included on every flight. They will, of course, have to be supplemented by the procedures and techniques in use by each particular crew. Some pilots, when flying in command, prefer to command every change in navigational frequencies. Other captains only want to be informed of the change. Some captains insist that certain routine items be taken care of by the copilot on his or her own initiative; other pilots shudder at this thought. The important fact is that the crew be totally and properly briefed before flight so that there is no uncertainty about the duties and responsibilities of each member.

CHECKLISTS

Their use and abuse

It would be extremely difficult to find a pilot who has not been exposed to a checklist at some point in time, probably very early in training. Unfortunately, it is not hard to find pilots who do not use or who misuse checklists. On the other end of the spectrum, pilots can be found who have checklists to manage their checklists. Neither extreme is ideal.

What they are and are not

A checklist is a tool used to initiate and direct effective cockpit communication and other activities pertinent to the conduct of the flight. It serves as a written aid in accomplishing the flight crew's duties and responsibilities, and it aims to ensure that the crew has properly configured the aircraft and its systems for the related phase of flight.

A checklist is NOT a substitute for a thorough knowledge of the aircraft and its systems and procedures. If all the information required to successfully fly an aircraft were contained in the checklists, training for type ratings would be easy. We could simply give a reading test, and when the determination was made that the applicant could read, the ground school would be considered complete. As anyone who has flown for more than a week knows, however, most things that happen in airplanes are not adequately covered by any checklist. The FARs require certain items to be covered by checklists, and in many aircraft, it is only those certain items that are found on the official checklists. Where have you ever seen a checklist that tells what to do if the oil pressure is high or the oil temperature low? Where is the guidance for unusual noises or strange smells? You won't find these answers in the checklist; they will only be found in your thorough understanding of the operation and integration of the aircraft systems.

Reasons for using checklists

There are many reasons for using checklists—almost as many as the excuses given for not using them—but some of the most important are:

- Checklists help to focus the crew's attention on the necessary procedures for the task at hand.
- Checklists eliminate guesswork by standardizing terminology and procedures.
- Checklists serve as an aid against failure of human memory, particularly during times of stress or high workload.
- Checklists help to balance the workload among the available pilots.
- Checklists help in avoiding aircraft limitations by ensuring that vital procedures are accomplished in the proper order.

Excuses given for not using checklists

There are many excuses that have been used by nonprofessional[1] pilots over the years for not using checklists. The excuses don't hold water, and they signal a dangerous attitude for one flying an aircraft, but let's list a few of these myths and debunk them right now:

- "I fly a simple aircraft. I don't need a checklist." What does the complexity of the aircraft have to do with the fallibility of human memory? Simple aircraft do not always have simple procedures, and stress and divided attention will make the best pilot forgetful.
- "I know my airplane so well that I have the checklist memorized, and I don't need it anymore." Oh, really! Again, memory deteriorates in times of stress, and the pilot is not usually aware of the deterioration. Additionally, it is unlikely that all of the items on a checklist will be properly recalled, even in times where the stress level is low.
- "Checklists take too much time." How so? Don't you have to accomplish all the same items whether or not the checklist is used? The same switches, levers, knobs, buttons, and dials have to be adjusted and checked with or without a checklist. The only extra time involved is that required to doublecheck that all is properly done, which is a wise investment.
- "Checklists are burdensome and increase pilot workload." Just the opposite is true. Checklists reduce cockpit workload by taking care of the organization and prioritization of necessary tasks.
- "Only beginners need checklists. I'm an ATP." Hmmmm. I wonder how the chief pilot of a major airline would respond to that statement?

Missed checklist items

Even with conscientious use of checklists, items are often missed. It is instructive to look at the reasons these items are missed and the procedures available to avoid the serious consequences of missing crucial items.

Workload intensity is a major reason for missed checklist items. This is almost always the result of trying to do too many things at one time. One person cannot listen to ATIS, contact approach, contact company for ETA and gate information, and compute landing-card information, plus accomplish the descent, in-range, and approach checklists just prior to intercepting the localizer and not be prone to rush and miss some items. The crew is usually its own worst enemy in this regard by delaying checklists to the last possible moment. To avoid this problem, checklists should be initiated and completed as early as is consistent with the phase of flight.

Complacency is all too often the cause for missing checklist items. Repetitive tasks are often performed without thinking about what is being done or how well it is being done. I have seen crews call out a checklist item and respond to its completeness with nothing having been done. Even the most experienced and conscientious crews are subject to this oversight.

Possibly the most common cause for missed checklist items is distractions; ATC is the most common cockpit distraction. When use of a checklist is most intense, the radio calls are also the most frequent. Unless there is a positive method of marking items completed or items undone, it is wise to redo an entire checklist when an interruption to the flow has occurred. This is probably the only sure method to prevent missed items.

Human memory fallibility

Human memory is fallible. Forgetting to enter a check in your checkbook register can be embarrassing. Forgetting a birthday or anniversary can lead to strained relationships. Forgetting to put the gear down can have far more serious consequences. An understanding of how memory works, and doesn't work, might help you understand what causes those gaps, oversights, and lapses that result in forgetfulness.

Each person's memory is made up of two components: short-term memory and long-term memory. Short-term memory is the doorway through which all information enters. Rehearsal of the items in short-term memory, such as repeating a phone number you wish to memorize, transfers information to long-term memory from which it can be recalled for use at later periods.

Short-term memory is extremely volatile and is capable of retaining about seven items[2] for about 15 seconds. Without rehearsal, this information is rapidly forgotten and overwritten by new incoming material. Because most interruptions to checklists come in the form of radio calls and traffic watches and not only are longer than 15 seconds but also insert new memory into short-term memory, the likelihood of accurately remembering and returning to the last checklist item completed is low.

Memory is subject to significant limitations that can make it unreliable and incomplete at times when it is most relied upon. Panic, anxiety, pressure, and confusion can all interfere with the ability to recall memorized items. Fatigue, illness, and hypoxia all tend to slow our ability to remember accurately. And items that are infrequently used are harder to recall in even the best of conditions.

A phenomenon known as interference can cause the pilot to forget or distort recalled information. This happens when two procedures are very similar but not identical. There is a tendency to combine or confuse the procedures during recall. Interference can also become a factor when there are many procedures, even though differing in important aspects, that were learned at about the same time, as in typical training situations. Use of checklists is an effective weapon against this sabotage of long-term memory.

Checklist design

Checklists come in all sizes and shapes. I have had the opportunity to view the homemade checklist one overly conscientious student of mine constructed for the normal operation of a Cessna 172. It was many times the length and weight of checklists used to fly 400-passenger airliners, and this particular pilot had a whole flight bag full of similar lists. One wonders when there was time to fly the airplane and watch for traffic between readings of this lengthy tome.[3]

The fact that an aircraft manufacturer has published a checklist does not mean that such a list is infallible or even correct. The taxi checklist for a popular corporate jet indicates that the trim should be set for takeoff, the controls checked, and then the autopilot should be checked. All true, but how many times have you checked an autopilot for operation and noticed that it moves the elevator trim? Quite often, I would venture to say, yet nowhere does the checklist call for a check or recheck of takeoff trim after the autopilot check. Most airlines publish their own checklists based upon the manufacturer's checklists but modified to provide better flow and logical grouping of items.

FAR 25, which governs certification of transport-category aircraft, lists the operating procedures that must be contained in the aircraft flight manual:

25.1585 Operating procedures.

(a) Information and instructions regarding the peculiarities of normal operations (including starting and warming the engines, taxiing, operation of wing flaps, landing gear, and the automatic pilot) must be furnished, together with recommended procedures for -

(1) Engine failure (including minimum speeds, trim, operation of the remaining engines, and operation of flaps);

(2) Stopping the rotation of propellers in flight;

(3) Restarting turbine engines in flight (including the effects of altitude);

(4) Fire, decompression, and similar emergencies;

(5) Ditching (including the procedures based on the requirements of 25.801, 25.807(d), 25.1411, and 25.1415(a) through (e));

(6) Use of ice protection equipment;

(7) Use of fuel jettisoning equipment, including any operating precautions relevant to the use of the system;

(8) Operation in turbulence for turbine powered airplanes (including recommended turbulence penetration airspeeds, flight peculiarities, and special control instructions);

(9) Restoring a deployed thrust reverser intended for ground operation only to the forward thrust position in flight or continuing flight and landing with the thrust reverser in any position except forward thrust; and

(10) Disconnecting the battery from its charging source, if compliance is shown with 25.1353(c)(6)(ii) or (c)(6)(iii).

(b) Information identifying each operating condition in which the fuel system independence prescribed in 25.953 is necessary for safety must be furnished, together with instructions for placing the fuel system in a configuration used to show compliance with that section.

(c) The buffet onset envelopes determined under 25.251 must be furnished. The buffet onset envelopes presented may reflect the center of gravity at which the airplane is normally loaded during cruise if corrections for the effect of different center of gravity locations are furnished.

(d) Information must be furnished which indicates that when the fuel quantity indicator reads "zero" in level flight, any fuel remaining in the fuel tank cannot be used safely in flight.

(e) Information on the total quantity of usable fuel for each fuel tank must be furnished.

FAR 91.503 sets forth the requirement for the availability and use of cockpit checklists:

91.503 Flying equipment and operating information.

(a) The pilot in command of an airplane shall ensure that the following flying equipment and aeronautical charts and data, in current and appropriate form, are accessible for each flight at the pilot station of the airplane:

. . .

(2) A cockpit checklist containing the procedures required by paragraph (b) of this section.

. . .

(b) Each cockpit checklist must contain the following procedures and shall be used by the flight crew members when operating the airplane:

(1) Before starting engines.

(2) Before takeoff.

(3) Cruise.

(4) Before landing.

(5) After landing.

(6) Stopping engines.

(7) Emergencies.

(c) Each emergency cockpit checklist procedure required by paragraph (b)(7) of this section must contain the following procedures, as appropriate:

(1) Emergency operation of fuel, hydraulic, electrical, and mechanical systems.

(2) Emergency operation of instruments and controls.

(3) Engine inoperative procedures.

(4) Any other procedures necessary for safety.

(d) The equipment, charts, and data prescribed in this section shall be used by the pilot in command and other members of the flight crew, when pertinent.

Aircraft checklists are organized into logical categories that have been proven over many years of use to provide for effective means for finding and accomplishing required procedures. These categories traditionally include normal, abnormal, and emergency checklists. Normal checklist procedures are the most routine, and these checklist items are the ones most often skipped. These checklists include procedures for the safe and efficient operation of the aircraft systems and equipment under normal circumstances. Gear-up landings, fuel-starved engines, and incorrect flap settings area all found in accident/incident reports, and all are preventable by adherence to checklists.

Abnormal procedures are usually those that deal with malfunctions or non-standard conditions that do not pose an immediate threat to the safety of the aircraft or its occupants. These would include such items as crossfeed procedures, failure of some portion of the electrical system, autopilot failures, wing-flap failures, and the like.

Emergency checklist procedures are used to isolate and compensate for failed or malfunctioning systems and are necessary to protect the aircraft and/or occupants from imminent serious or critical harm. Emergency procedures and checklists are often associated with the illumination of a red warning light in the cockpit and include such items as engine fires and failures, electrical fires, explosive or rapid decompression, and other events of a serious and/or threatening nature.

A well-designed checklist should contain only those items necessary to accomplish the task at hand. Extra and irrelevant items encourage pilots to bypass the use of the checklist. A good checklist should also have the items listed in the correct order and be arranged so as not to interfere with flying the aircraft.

Checklist procedures—challenge and response

The challenge-and-response system of checklist management has been used for decades by professional pilots to ensure good communication and standard procedures when dealing with normal, abnormal, and emergency situations. It is a proven method of checks and balances that helps to ensure that all checklist items and all checklists are accomplished correctly.

Challenge and response is a technique in which a pilot initiates a checklist item by saying it out loud exactly as it is written: a challenge. The pilot responsible for performing the item indicates its accomplishment by the designated response. This system, when used consistently, is effective no matter whether two pilots are involved or one pilot is responsible for both the challenge and the response. Company policy, aircraft flight manuals, and mode of flight will determine whether one pilot will be responsible for both challenge and response or whether both pilots will be involved.

Besides the obvious benefit of organizing the cockpit tasks and ensuring their timely completion, the challenge-and-response system has a benefit not often appreciated. The simple act of saying the checklist items out loud stimulates the sense of hearing and not only helps focus the attention but also increases the amount of attention the pilot can devote to a checklist item. The verbalization helps both pilots to monitor the checklist progression and helps to prevent confusion and distraction in high-stress situations.

SITUATIONAL AWARENESS

Situational awareness is the accurate perception of the factors and conditions that affect the aircraft and the flight crew during a specific period of time. Simply put, it is knowing where you are and what is going on around you at all times. There is a very direct relationship between situational awareness and safety. Pilots who have a high degree of situational awareness are less prone to make the common types of errors that make headlines.

Because part of the definition of situational awareness is the term "perception," it is very likely that situational awareness among members of a group will vary as their perceptions of the factors and conditions affecting the flight will vary. Various members of a group might have high situational awareness or low situational awareness, but one thing is certain, there will be variance. Safety of flight, however, depends on the situational awareness of the group as a whole. This group situational awareness is not the sum total of the individual pilot's level of situational awareness. It is, rather, limited to that level of situational awareness achieved by the individual in command. Unfortunately, this dynamic has led to a number of preventable fatal accidents. The captain of the Air Florida 737 that crashed into the 14th Street Bridge in Washington, D.C., did not have the level of situational awareness attained by the copilot. Unfortunately, the copilot was not able to raise the captain's level of awareness in time to save the flight.[4]

In order to raise the crew's level of group awareness, it is necessary for the copilot to contribute his or her situational awareness to the captain, and it is necessary for the captain to accept it. This can be a sticky situation, to say the least. Many airlines have given their flight pilots training in assertiveness in addition to workshops on effective communication and leadership. This cooperation and sharing of awareness is the key to achieving high group situational awareness. Without better situational awareness, the flight is in jeopardy.

Elements of situational awareness

Flying skills Remember how much effort and attention it took just to maintain straight-and-level flight in those early student days? Little time and attention remained for navigation, traffic watch, or troubleshooting any system malfunctions. Luckily for most of us, as we progressed in our flying careers, our physical flying skills improved tremendously, and consequently more time and attention can be devoted to the mental aspects of flying.

It should be pointed out, however, that physical flying skills are extremely important and no pilot should complacently believe that that stage of learning and improvement is in the past. We all need continual training and practice to remain at the top of our potential physical performance. And contrary to popular opinion, practice does NOT make perfect. Practice only makes permanent. If you would strive for perfection, you must make your practice as close to perfection as possible. My instructor once told me that the day would come in my flying career that maintaining an exact altitude would make the difference between life and death. He emphasized that when that day did arrive, maintaining that exact altitude had best be automatic because there were going to be many other things competing for my attention. The day has not yet come, but I have no reason to doubt his word.[5]

Experience and training Experience and training are closely related. Training is an experience and experience gives us training in day-to-day realities. Both experience and training add to our mental file of facts, which is constantly updated and consulted as new situations arise. We draw upon our experiences to establish how conditions and events are interpreted and how we respond to them. Our training and experience allows us to solve problems more quickly and therefore to devote more time to other problems and situations requiring attention.

Most problems faced by pilots are solved before the flight begins. Through adequate and regular training, the procedures associated with normal and emergency situations are filed in our mental file just awaiting retrieval at the appropriate time. Reaction time and attention required to respond to the emergency are consequently reduced.

Spatial orientation Spatial orientation goes somewhat further than knowing what seat you are occupying. It is more than simply knowing the aircraft's attitude, heading, altitude, configuration, and airspeed. Spatial orientation includes an accurate perception of the aircraft's position in relation to navigational aids, airports, terrain, weather, and other aircraft. It is a complete mental picture of where we are, where we are going, and when we are going to get there.[6]

Health and attitude Health and attitude are key elements in the situational awareness equation, but they are seldom given the importance due them. Physical illness directly affects our perception and interpretation of events in our environment. Emotional state and attitude directly affect a pilot's ability to achieve and maintain situational awareness. Preoccupation with personal problems takes attention "units" that can better be devoted to the safety of flight. A good, positive, and professional attitude allows the pilot to focus abilities and energies on the job at hand.

Detecting loss of situational awareness

Perhaps one of the most important factors concerning the loss of situational awareness is that in some ways it mimics hypoxia (which, by the way can certainly be a contributing factor) in that this loss can be insidious and unnoticed by the pilot until the situation is almost beyond recovery. If such is true, what clues can be watched for to alert the crew that situational awareness is but a memory? Among the clues to possible or actual loss of situational awareness are the following:

- Ambiguity. Two or more independent sources of information are in disagreement. These two sources might be two or more independent indicators for the same aircraft system (projected position and actual position, pilot and copilot, etc.). As long as the conflict is unrecognized or unresolved, situational awareness is in jeopardy.

- Fixation or preoccupation. A pilot in this situation is clearly not paying attention to the task at hand and loss of situational awareness is obvious, at least to outside observers. The pilot so affected will rarely, if ever, recognize this loss in himself or herself and it is up to other pilots to observe and alert the preoccupied pilot.

- Confusion. Confusion as to clearances, procedures required, weather options, and other flight information is a good indicator that situational awareness is in serious jeopardy, if in fact not already lost.

- Nobody flying the airplane. Even with the autopilot engaged, it is necessary that one pilot be in command of the aircraft and responsible for monitoring progress both inside and outside the aircraft. Without such monitoring, unintended and uncorrected deviations are a real possibility, and situational awareness has been seriously compromised.

- Improper procedures. Anytime consideration is given to the use of improper or nonstandard procedures, in the absence of overriding reasons, the possibility of loss of situational awareness exists. Standard operating procedures and associated limitations are established to ensure that actions are taken in the proper sequence and that human fallibility does not overreach the capabilities of the pilot and aircraft. Improper or nonstandard procedures put the flight outside of what has been established as the safe-flight operating envelope.

Whenever either pilot becomes aware that any of the above have taken place, an assessment of the flight conditions should be immediately made and steps taken to return the crew to a high level of situational awareness and safety.

MANAGING COCKPIT SAFETY

"And they called it pilot error. . . ."—the catchall phrase that haunts our worst nightmares. A phrase we hope to never have ascribed to our actions, and yet, how do we guard against this, the worst and final verdict?

With modern technology and reliability of aircraft and systems, fewer aviation accidents and incidents are being ascribed to mechanical failure and more and more are being ascribed to pilot error. Is this an accurate assessment of the cause, or should it more realistically be termed "human factors"? Many dangers lie in wait for the unwary crew. The cockpit is a workplace that requires that the workers employed there possess a high degree of skill and training, but all too often that training stops just short of the solution to the major risk to flight. Crews are rigorously trained and drilled in the simulator until they can repeat aircraft emergency procedures in their sleep. When quizzed, they can discuss and describe aircraft systems in great detail.[7] When quizzed about effective cockpit management, however, many have little to say.

NASA, FAA, and NTSB have been involved in studies of crew interaction and the dynamics of cockpit safety management. NASA has generated a list of the most common areas of flight crew failures related to aviation safety:

- Preoccupation with minor mechanical problems
- Inadequate leadership
- Failure to delegate tasks and assign responsibilities
- Failure to set priorities
- Inadequate monitoring
- Failure to utilize available data
- Failure to communicate intent and plans

In the study and identification of the most common failures, NASA has also generated six operating rules for improvement in cockpit-safety management:

- The positive delegation of flying and monitoring responsibilities must be the top priority in both normal and abnormal situations. As noted, when nobody is flying or monitoring, there is a serious loss of situational awareness. It cannot just be assumed that the pilot who made the takeoff is the pilot responsible for flying the aircraft. Note the word "positive" in the first sentence. The delegation of responsibilities must be verbalized and acknowledged by both pilots.

- A positive delegation of monitoring responsibilities is just as important to safety as a positive delegation of flying responsibilities. It is common to hear pilots make a definitive and positive statement as to who is flying: "I've got the controls" acknowledged by the other pilot's statement "You've got it." Far less common is the acknowledgment as to who is doing the monitoring. This, too, needs to be verbalized and acknowledged. Never assume anything in the cockpit of an aircraft. It must be stated and acknowledged which pilot is responsible for monitoring. This monitoring is all inclusive. Everything that is done to accomplish the flight mission is monitored: engine performance, performance versus flight plan, navigation, weather, and the like.

- The pilot flying the aircraft should avoid performing secondary tasks during dynamic phases of flight, unless necessary. Some captains get impatient and accomplish items (usually of the "housekeeping" type) rather than wait for the other pilot to accomplish them in turn. This not only is destructive to situational awareness, but it takes the pilot's attention away from his or her major responsibility of flying the airplane. This is a disruptive practice anytime during the flight, but during dynamic situations (climbs, descents, course changes, etc.), it can be particularly dangerous. This attitude is also further destructive of the working relationship that must exist if safety is to be enhanced.

- Whenever there are conflicting interpretations of fact, external sources of information must be used to resolve the conflict. We have already discussed the concept that existence of ambiguity can be taken as an indicator of the breakdown of situational awareness. Perhaps that should be clarified somewhat to state that the persistence of an ambiguity can be taken as an indicator of the breakdown of situational awareness. If an ATC clearance is interpreted by one pilot to mean one thing and by the other pilot to mean another, there is a conflict. Consult the controller. If there is a conflict about the management of aircraft systems, consult the aircraft flight manual. A difference of opinion is not a serious problem. Attempting to solve that difference or ambiguity with "confidence" in the opinion of one pilot or the other is serious. Consult an authoritative outside source. (One pilot should be consulting the outside source while the other pilot continues to fly the airplane.)

- Whenever there is conflicting information from two sources or information of questionable validity from one source, crosschecking from an independent source is required. A hypothetical but illustrative example is the disagreement of two independent navigation sources as to present position. NAV 1 says you are left of course; NAV 2 insists that the course is to the left. There are mountains to the right and left of course.[8] This is no time to be deciding which navigation receiver you place the most confidence in. Get a third opinion. Call ATC or consult your brand-new whiz-bang hand-held GPS for a correct position.[9]

- If any pilot has a doubt about a clearance, procedure, or situation, he or she must make that doubt known to other pilot. This perhaps can be the most difficult rule for a new copilot to master. There is the old persisting concept that the captain is the only authority in the cockpit. While it should be emphasized that only one person can be in command and be of ultimate authority, that command authority does not make the captain infallible. There is more than one case of a flight that encountered difficulties because of the unexpressed doubts of a copilot. Recall that preoccupation is one indicator that situational awareness is lost. Timely expression of doubt about the action, or lack of action, of the other pilot can save the day and help to reestablish situational awareness.

Thoughtful consideration of these rules will undoubtedly raise more questions, and that is as it should be. The pursuit of aviation safety should be a foremost topic for all who venture into the skies, whether for business or pleasure. In addition to the above recommendations of NASA, there are five skills that every pilot needs to develop to maximize his or her individual contribution to aviation safety, even if he or she never flies with another pilot:

- Asking the right questions. Data collection and interpretation is an ongoing process during the planning and conduct of every flight. The important skill involved here is to be assertive in obtaining *all* of the available information. Pilots must ask each other, ask air traffic control, ask flight service, tune all available navaids, and use any and all sources likely to provide the answers needed for the safe conduct of the flight. Equally important, however, is the critical evaluation of the validity of the answers. It is dangerous to put blind faith in the decisions or opinions of others, no matter their position or experience. Eliminating the crucial evaluation of the soundness and objectivity of the answers received can result in a situation that is as bad as one created by no communication at all.

- Frankly stating opinions. Your convictions and opinions are important. It is your professional responsibility to state facts or insights you have that you feel are valuable to the safe conduct of the flight. It is also your responsibility to listen to the opinions of others and be prepared to change your opinions when the circumstances and objective facts presented warrant such a change. A forthright statement of your opinion might be the last chance to transform a dangerous situation into a safe one. (Examples: "We are below the MSA for this area." "If we remain in this holding pattern, I believe we will be compromising reserve fuel.")

- Working out differences. Harmoniously resolving the inevitable differences that occur during day-to-day operations is a crucial element in the cockpit-safety-management equation.

- Making decisions. Questions may be asked, opinions stated, differences worked out, but inevitably a decision must be made as to a course of action. It should be kept in mind that not making a decision is a wrong decision. Decisions must be made and acted upon in a timely manner to assure the safety of flight.

- Criticizing constructively. This is the skill that is best practiced after the flight when time has permitted an evaluation of the decisions made and actions taken. The ability to criticize constructively and to take constructive criticism with an open mind is the catalyst that turns simple statements into avenues for discussion to work out differences and solve problems.

When these five skills are developed and used together, they provide a solid framework for display of physical flying skills. These skills, like any other, require constant practice to keep them current and usable. A pilot with a high degree of proficiency in these skills, and a high degree of awareness of the hazards of hypoxia, will be an unlikely candidate for an accident resulting from human-factors problems: "pilot error."

PILOT INCAPACITATION

Pilot incapacitation comes in two distinct varieties: obvious and subtle. Of the two, the subtle variety is by far the most dangerous since it tends to exist for a much longer time before it is recognized and corrected. Few aircraft accidents have been caused by a pilot slumping over unconscious at the controls. Many accidents have been caused by the body being present but the mind being absent.

Obvious incapacitation

Obvious pilot incapacitation means just what it says—it is obvious to all present that a pilot has become incapacitated. It will normally not be possible to determine the reason for incapacitation until medical help is obtained, but there are some definite steps to be taken by the remaining crew and/or passengers to deal with the situation:

1. Have the affected pilot removed from his or her seat if at all possible. If not possible to remove him or her, attempt to restrain the pilot or secure the pilot's hands. If the incapacitated pilot begins to regain consciousness, there will be a period of agitation, confusion, and possible combativeness prior to higher mental functioning returning. This is not the type person you would want to have sitting in the other seat when you are making your first solo flight in a jet.

2. Confess your predicament to ATC. Not only can a controller give you expedited handling, but he or she can remove some of your workload through vectors and information providing. ATC will also be able to summon medical help to meet your flight.

3. Enlist the aid of other pilots or passengers. Many corporations are now providing training for those who routinely fly in company aircraft to assist the crew in the event of incapacitation emergencies. Even without specific training, passengers can read checklists, retrieve approach charts, and accomplish other tasks not requiring a high degree of aviation training.

4. Reorganize your cockpit and prioritize your duties. You might want to change seats if the incapacitated pilot has been removed from the cockpit. Fly from the seat most comfortable to you and giving you the best access to the controls and switches needed to complete the flight. Make the decision early as to where you are to land the airplane. Your destination might not be the place to go even if it is the closest airport in point of distance. Perhaps a larger, better-equipped airport that is more convenient to emergency medical facilities will be more appropriate.

Keep in mind, however, that your PRIME responsibility is the safety of the flight, not the continued health of the incapacitated pilot. He or she might be your best buddy, but putting the flight in jeopardy over concern for that person's health serves no one. Your job now is to safely land the aircraft, THEN worry about your compatriot. Select the destination that provides the most optimal conditions for approach and landing for you, operating as a single pilot, without consideration for anything but the safety of the flight. Don't flirt with low weather, high cross-winds, or other similar situations if avoidable.

Subtle incapacitation

Have you ever been driving your car and suddenly arrive at an intersection or landmark and not remember passing other familiar places en route? To the outside observer, you might appear to be paying attention to your driving, but your mind is not on the business at hand, and you are a danger to yourself and those around you. We have all experienced this, and it is a form of subtle incapacitation. A pilot, too, might appear to be functioning but be suffering from a subtle form of inca-pacitation.

Because the pilot might appear to be fine, how will we determine that he or she is, in fact, disabled? The best way to guard against this hazard is through stan-dard procedures, and in particular standard callouts and challenge-and-response checklists. As was stated before, each of the standard callouts should generate a re-sponse. In the event that the other pilot does not make the proper response by the second call, consider that pilot to be disabled, and take the appropriate action. This can be a touchy situation, but the simple phrase "I've got the controls" should be the immediate action. In all but the worst cases, this will likely yield a response from the other pilot. It should be determined at that time whether the lapse was momentary or more persistent.

This can be called, for want of a better name, the "Art of Diplomatic Mutiny." One pilot must be in control of the aircraft at all times, and during crucial phases of flight, there is little time to discuss the matter. It should be a matter of training and agreement among pilots exactly what the criteria for assuming control should be, but once these criteria are agreed upon, they should be adhered to. No pilot should allow another to command an aircraft while in an incapacitated condition. The fear of reprisals once the flight has been completed should not influence the inflight decision if the criteria for change in command are met. Think about it. Would you really allow someone to endanger your life out of fear of being severely reprimanded?

Causes of incapacitation

Hypoxia I cannot overemphasize the seriousness of this risk. Every pilot's reactions to oxygen deprivation will vary on a day-to-day basis. (See chapter 3 for more information.) Be aware also that hypoxia has been known to trigger symp-toms in someone with a history of brain pathology.

Hypoglycemia Pilots are perhaps the most erratically nourished group in the world. Many of us have become more health conscious in recent years and when at home carefully watch our diet. On flying days, however, the opposite might be true. Hastily downed cups of coffee and a candy bar for quick energy puts us on a blood-sugar roller coaster that has a definite affect on performance. Low blood sugar can leave a pilot feeling headachy, tired, and irritable. Concentration becomes difficult, and errors are more easily made. Try to have healthy snacks handy that are high in protein to sustain your blood sugar over a much longer period.

Dehydration The climate in the cabin is drier than the driest desert. Dehydration can leave you feeling headachy, dizzy, and fatigued. Drink PLENTY of water while flying. The body loses about a quart of water a day through perspiration and respiration. This is under ideal conditions without strenuous exercise. The dry climate in the cockpit will accelerate this loss. Unfortunately, the beverage of choice for many pilots is a cup of coffee, which is probably one of the worst things for someone trying to combat dehydration. Coffee is a *diuretic*, which is a substance that accelerates removal of water from the body. Lack of thirst is not a good indicator of the state of hydration of the body since thirst does not become apparent until dehydration is already present. Drink plenty of water throughout flight to combat this risk.

Previously unrecognized medical problems Unrecognized illness or the early stages of an illness can sap energy and make a pilot more susceptible to hypoxia and other stresses of flight. This can lead to a lack of concentration and perception.

Personal problems The past-due bill, the fight with the significant other, the leaky roof: The list can go on and on, but suffice it to say that many of the stresses and occurrences of everyday life can impinge upon our consciousness when flying and sap our concentration and take attention away from the task at hand.

In summary, cockpit resource management is a dynamic process involving the entire crew and using all information, aids, and procedures to ensure the safe and efficient outcome of the flight mission. No single person is unimportant in this equation.

Endnotes

1. Remember, professional does not refer to the author of your paycheck. Professionalism is an attitude, not a job title.

2. It's no coincidence that telephone numbers in the United States are seven digits long.

3. This particular pilot carried checklist use to an extreme and was totally compulsive about usage of his checklist. We often commented that if he ever found himself on collision course with a 747 he would die because he would never be able to locate his "AVOID 747" checklist in time to avert disaster. His approach to checklist usage was just as dangerous as the "I-don't-need-a-checklist" attitude.

4. The copilot had indicated to the captain on more than one occasion during the takeoff roll that the power setting did not seem to make sense. In fact, the temperature probe in the engine inlet was iced over giving an erroneously high EPR reading. The captain chose to ignore these warnings.

5. You see, Mr. Dwyer, I *did* listen to what you said!

6. I tell my beginning instrument students that a good instrument pilot knows three things: (1) where the closest VFR weather is, (2) where the lower terrain is,

and (3) where the wind is. These are all, of course, predicated upon constantly knowing exactly where you are!

7. Probably in too great detail: The past (and I truly hope it is past) emphasis on an engineeringlike knowledge of aircraft systems placed too great a value on the memorization of trivia. What enhances the safety of flight to know that the fusible checkvalves in the fire-extinguishing system of an aircraft melt at 205–210°F or 2400 PSI and discharge the contents into the tailcone? This cannot be detected in flight, and should you be psychic and detect it, nothing can be done from the cockpit to correct or change the situation. The fire bottles are inspected on preflight, and their pressure compared with a table of ambient temperatures to determine that the pressure indicated is within limits. Memorization of any other trivia is unnecessary and useless and the time spent in that pursuit can better be spent in understanding total systems integration and management. This was, however, an actual question proposed by the FAA for an equipment oral.

8. Check the VOR/DME approach to Ely, Nevada. The approach course follows a valley. After executing this approach in VFR conditions, I doubt if I would feel totally comfortable flying it in the clouds!

9. No, I am not advocating navigation by reference to a navigation radio not installed in the aircraft and not certified for IFR en route navigation. These receivers, like hand-held communications radios, are nifty little emergency devices, however.

Glossary

THE ENTRIES IN THIS GLOSSARY COME NOT ONLY FROM THE PRECEDING text but are words and terms useful in the overall understanding of the operation of jet aircraft.

accelerate-stop distance—The length of runway required for an aircraft to accelerate to a specified speed, usually V_1, and then be able to stop on the runway.

acceleration check—A maintenance check of a gas turbine engine in which the time required for the engine to accelerate from idle RPM to its rated-power RPM is compared with the time specified for this acceleration by the engine manufacturer.

accelerometer—A sensitive instrument that measures the amount of force exerted on an object because of its acceleration.

accessory drive shaft—The shaft, driven by bevel gears from the compressor shaft, which drives the accessory gearbox.

accessory gearbox—A part of a gas turbine engine that contains the drive gears to operate such engine-driven accessories as fuel pumps, oil pumps, generators, tachometer, and the like.

adiabatic lapse rate—The rate at which air cools as it is forced upward or warms as it sinks if no heat energy is added to it and none is taken from it. Under standard conditions, the adiabatic lapse rate of dry air is 3°C or 5.4°F per thousand feet.

adiabatic process—The process by which fixed relationships are maintained during changes in temperature, volume, and pressure in a body of air when heat is neither added nor removed from the air.

aerodynamic center—The point along the chord of an airfoil where all changes in lift effectively take place. The aerodynamic center is not affected by the camber or thickness of the airfoil nor by the angle of attack. For a subsonic airfoil, the aerodynamic center is located between 23 percent and 27 percent of the chord length back from the leading edge of the chord, and for a supersonic airfoil, it is located 50 percent back from the leading edge of the chord.

aerodynamic force—The net resulting static pressure multiplied by the planform area of an airfoil.

afterburner—A device in the exhaust system of a jet engine used to increase the thrust of the engine. Because most of the air passing through a gas turbine engine is used for cooling air, the exhaust contains a large proportion of oxygen that can be mixed with additional fuel to create additional thrust. Fuel is sprayed into the hot, oxygen-rich exhaust where it burns.

air brake—A device that can be extended from the structure of an airplane to produce a large amount of parasite drag and allow the aircraft to increase its angle of descent without increasing speed. Air brakes are also called *speed brakes* or *dive brakes* and differ from flaps in that they produce no lift.

air-cooled turbine blades—Hollow blades in the turbine wheel of gas turbine engines cooled by compressor bleed air passing through them.

air-oil separator—A device in the vent portion of the lubrication system of gas turbine engines that separates the oil from the air before venting the air overboard.

angle of attack (AOA)—The acute angle between the relative wind and the chordline of an airfoil.

annular combustor—A single-piece combustor for gas turbine engines made in the shape of a ring or donut. Annular combustors make the most efficient use of the space they occupy and are the most efficient type of combustor used in gas turbine engines.

annunciator panel—A panel of warning and advisory lights in plain view of the pilot of an aircraft. Red lights are used to indicate a dangerous condition and usually require immediate action. Amber lights show system malfunctions of a less serious nature.

anti-icing fluid—A fluid, usually made of some form of alcohol and glycerin, that is sprayed on an aircraft to keep ice from forming on the surface.

anti-icing—The prevention of the formation of ice on a surface.

antiservo tab—An adjustable tab attached to the trailing edge of an aircraft control surface. The antiservo tab moves in the same direction as the control surface and produces a stabilizing aerodynamic force that counters some of the aerodynamic force that moves the surface.

antiskid system—A system of control of aircraft brakes that releases brake pressure in the event the wheel begins to lock or skip. This can occur several times per second.

APU—Auxiliary power unit. A self-contained engine generator carried aboard an aircraft to generate power for ground operation and engine starts.

axial loading—An aerodynamic force that tries to move the compressor forward in its case. Axial loading is supported in a gas turbine engine by ball bearings.

axial-centrifugal compressor—A combination axial-flow and centrifugal-flow compressor used in a gas turbine engine. The axial-flow portion serves as the low-pressure compressor and the centrifugal-flow portion serves as the high-pressure compressor.

axial-flow compressor—A type of compressor found in gas turbine engines in which the air passes through the compressor parallel to the axis of rotation of the compressor. Axial-flow compressors are typically made up of several stages of alternating rotors and stators.

azimuth—Angular measurement made in a horizontal plane and in a clockwise direction. Refers to compass direction.

balance tab—A small tab on a primary control surface of an airplane that moves in the direction opposite that of the control surface to reduce the amount of force needed by the pilot to move the surface.

balanced control surface—A movable primary control surface of an airplane designed in such a way that the aerodynamic forces help the pilot move the surface.

ball bearing—A type of antifriction bearing used to support thrust loads as well as radial loads. A ball bearing consists of outer race of hardened steel with a groove in it mounted in a stationary housing and an inner race with a matching groove mounted on a rotating shaft. Polished, hardened steel balls ride in the grooves and support the shaft.

bar—Unit of absolute pressure equal to 1 million dynes per square centimeter. The millibar, one-thousandth of a bar, is a unit of absolute pressure used in meteorology. The standard absolute pressure at sea level is 1013.2 millibars.

bleed valve—A valve in the compressor case of a gas turbine engine that allows some of the air in the higher stages of the compressor to be bled off, thereby improving the airflow through the engine and lessening the danger of compressor stall.

boundary-layer control—A method of decreasing aerodynamic drag caused by turbulent flow of the boundary layer by using either a high-velocity blast of air to blow the random flowing air off the surface or low pressure below the surface to suck the boundary layer air off the surface through tiny holes or slots.

boundary layer—The layer of air that flows next to an aerodynamic surface.

bow wave—A shock wave that forms when an aircraft is flying at a speed faster than the speed of sound. A bow wave forms either immediately in front of the aircraft or is attached to the nose of the aircraft.

Boyle's law—Law of gases that states that the product found by multiplying the pressure of a gas by its volume will always be constant.

Brayton cycle—The constant pressure cycle of energy release used to describe the action of a gas turbine engine. All events (compression, combustion, expansion, and exhaust) take place at the same time but in different parts of the engine.

buffet—Turbulent movement of air over an aerodynamic surface.

bug—Movable marker on a flight instrument that can be set to reference a particular indication.

bug speed—Common term for V_{REF}.

burst RPM—Compressor speed of an engine at which the centrifugal loads are so high that the engine will fly apart.

bypass ratio—The ratio of mass airflow in pounds per second through the fan section of a turbofan engine to the mass airflow that passes through the gas-generator portion of the engine.

camber—The amount of curve of an airfoil section.

carbon seal—A seal in a gas turbine engine between a rotating shaft and a fixed housing designed to keep oil from flowing into the gas path of the engine.

cartridge-pneumatic starter—A pneumatic starter for a gas turbine engine that uses air pressure built up by powder in a cartridge in the breech of the starter.

cavitation—The formation of an area of low pressure behind an object moving in a fluid.

center of lift—The point along the chord line of an airfoil at which all the lift forces produced by the airfoil are considered to be concentrated.

center of pressure—The point on the chord line of an airfoil where all the aerodynamic forces are considered to be concentrated.

center of pressure—The point on the chord line of an airfoil through which the aerodynamic force acts.

centrifugal flow compressor—A compressor in a gas turbine engine in which air is taken into the center of the impeller and is slung outward by centrifugal force into the diffuser where its velocity is decreased and its pressure is increased.

Charles's law—A basic gas law that explains the relationship between the temperature of a gas and the volume of that gas. If the pressure of a gas is held constant and the temperature is increased, the volume will also increase.

chine tire—An aircraft nosewheel tire that has a deflector, or chine, molded into its sidewall designed to throw water or slush on the runway to the side so that none gets into the engines. Chine tires are used on jet aircraft with rear-fuselage-mounted engines.

chip detector—An electrical warning system that detects the presence of metal chips in the lubrication system of an engine.

choked nozzle—The exhaust nozzle of a gas turbine engine designed in such a way that the flow rate of the exhaust gases cannot exceed the speed of sound.

chord line—A straight line connecting the leading and trailing edges of an airfoil.

chord—The distance between the leading and trailing edges of an airfoil. The chord is the characteristic dimension of an airfoil.

clear air turbulence (CAT)—Extreme turbulence found at high altitude created when a difference in temperature causes a violent movement of air.

cold section—The portion of a gas turbine engine ahead of the combustors. The compressor is the main component of the cold section.

combustion liner—The perforated-steel inner liner in the combustion chamber of a gas turbine engine.

combustor efficiency—The ratio of the amount of heat energy released by burning the fuel in a gas turbine to the amount of heat energy contained in the fuel.

combustor—The section of a gas turbine engine in which the fuel and air are mixed and burned.

compressibility error—The airspeed indicator error caused by the compression of air at the forward end of the pitot tube.

compressor bleed air—Air bled off one of the stages of the compressor of a gas turbine engine.

compressor case—The outer housing of the compressor section in a gas turbine engine.

compressor pressure ratio—The ratio of the pressure of the air at the compressor discharge to the pressure of the air at the compressor inlet.

compressor stage—One compressor stage in a gas turbine engine is made up of one set of rotor blades and the following set of stator blades.

compressor stall—A condition existing inside a gas turbine engine when the angle of attack of the compressor blades becomes excessive.

compressor surge—A severe compressor stall across the entire face of the compressor. A compressor surge can cause a flameout, or, if severe enough, can cause structural damage.

compressor—The section of a gas turbine engine with components that increase the pressure of the air flowing through the engine.

conservation of energy—A physical law stating that energy cannot be created or destroyed but only be changed from one form into another.

constant-speed drive (CSD)—A drive unit used with gas turbine engines to couple ac generators to the engine. The CSD keeps the speed of the generator constant; thus, the frequency of the ac power generated remains constant as the engine speed varies throughout its operating range.

convergent duct—A duct through which subsonic air flows to increase its velocity. The cross-section of a convergent duct becomes smaller in the direction of the airflow.

core engine—The gas generator portion of a gas turbine engine.

creep—A condition of permanent elongation of a turbine blade in a gas turbine engine. Centrifugal loading acting on turbine blades at high temperature causes the blades to actually grow in length.

critical angle of attack—The highest angle of attack at which air will flow smoothly over an airfoil, also sometimes called the *stalling angle of attack*.

critical Mach number—The flight Mach number at which there is the first indication that air flowing over any part of the aircraft, usually the wing, has reached the speed of Mach 1, the local speed of sound.

cross-bleed—A method of starting gas turbine engines in which compressor bleed air from one engine can be directed to another engine where it can be used to operate an air starter.

cuff—Specially shaped piece of metal attached to the leading edge of a wing to increase the camber and improve the slow-flight characteristics of the wing.

customer bleed air—A bleed from a gas turbine engine used for some function other than the operation of the engine.

Dalton's law—A basic gas law that states there is always the same number of molecules of gas in a closed container as long as the gas is held at a constant temperature and pressure. Dalton's law explains the partial pressure of gases in the atmosphere.

deep cycle—Maintenance procedure required on nickel-cadmium batteries to equalize all the cells. The battery is completely discharged, allowed to stand in its shorted condition for a specified period of time, and then recharged to 140 percent of its ampere-hour capacity.

demand oxygen system —An oxygen system that meters oxygen from the regulator to the mask only when the user inhales.

diffuser—A duct installed at the output of a centrifugal compressor to reduce the velocity of the air leaving the compressor and thus increasing its pressure.

diluter-demand oxygen system—A form of demand oxygen system that dilutes the oxygen supplied to the masks with air from the cabin when the aircraft is flying at low altitude. The amount of cabin air metered into the masks decreases with altitude.

divergent duct—A duct with cross-sectional area that increases in the direction of fluid flow; the duct decreases velocity and increases pressure.

drag—The force, parallel to the relative wind, that opposes the motion of a body through the air.

duplex fuel nozzle—A type of fuel nozzle that discharges its fuel into the combustor at two different rates: one for low-power operations and the other for higher-power operations.

dynamic pressure—The pressure that a moving fluid would have if it were brought to a stop.

ejector—The part of a jet pump that produces low pressure as fluid flows through it.

engine-pressure ratio (EPR)—The ratio between compressor-inlet total pressure and turbine-discharge total pressure used as an indication of the amount of thrust being developed by an axial-flow gas turbine engine.

engine stations—A method of identifying the various locations in a gas turbine engine for ease of describing pressures and temperatures. According to station number:

1. Tip of nose cone
2. Low-pressure compressor inlet
3. High-pressure compressor inlet
4. Combustor inlet
5. High-pressure turbine inlet
6. Low-pressure turbine inlet
7. Turbine discharge
8. Turbine-duct discharge
9. Exhaust-nozzle discharge

engine trimming—Adjustment of the fuel-control unit of a gas turbine engine to bring the engine to the specific temperature, fuel flow, thrust, and pressure ratio recommended by the manufacturer.

entrained water—Water held in suspension in the fuel carried in an aircraft fuel tank.

EPR-rated engine—Gas turbine engine with a rated thrust expressed in terms of engine-pressure ratio.

exhaust nozzle—The opening in the tail pipe of a gas turbine engine through which exhaust gases leave the engine.

exit-guide vanes—Stationary airfoils at the discharge of an axial-flow compressor designed to straighten the airflow so that it leaves the compressor in an axial direction.

false start—A condition in which the fuel-air mixture inside a gas turbine engine ignites and burns but the RPM does not build up high enough for the engine to continue to run without the aid of the starter. Also called a *hung start*.

fence—A stationary vane that extends chordwise from the leading edge to the trailing edge of a wing to keep the air from flowing along the span of a wing.

flameout—A condition in the operation of gas turbine engines in which the fire in the engine unintentionally goes out.

flight idle—An engine speed that produces a minimum amount of flight thrust.

flutter—Rapid and uncontrollable oscillation of a flight-control surface of an aircraft caused by a dynamically unbalanced condition.

foot-pound—The amount of torque produced when 1 pound of force is applied 1 foot from the point of rotation. As a measurement of work, it is the amount of work done by a 1-pound force when it causes movement of 1 foot.

foreign object damage (FOD)—Damage to a gas turbine engine caused by an object being sucked into the engine while it is running.

free turbine—A turbine stage in a gas turbine engine that is not used to drive the compressor in the gas generator section of the engine.

gas generator—Basic gas turbine engine consisting of the compressor, diffuser, combustor, and turbine. Also called the *core engine*.

ground idle—The engine speed normally used for operating a gas turbine engine on the ground so that it produces the minimal amount of thrust.

heat exchanger—A device in which heat in one substance is transferred to another substance by conduction.

high-bypass turbofan engine—An engine in which the fan moves more than four times as much air as is moved by the core of the engine.

high-pressure compressor—The high-speed compressor in a two-spool, axial-flow gas turbine engine. Normally known as the N_2 compressor.

high-pressure turbine—The stage or stages of turbine used to drive the high-pressure compressor in a two-spool, axial-flow gas turbine engine.

hot section—The portion of a gas turbine engine that operates at high temperature including the combustors, the turbine, and the exhaust system.

hot start—A condition in which a gas turbine engine starts, but its internal temperature rises high enough to damage the engine.

hot-section inspection—An inspection of the hot section of a gas turbine engine.

hung start—See *false start*.

idle thrust—The thrust produced by a gas turbine engine when the power control lever is pulled back to the idle stop.

igniter—A component in a gas turbine engine used to ignite the fuel-air mixture when the engine is being started.

inlet guide vanes—Stationary vanes located in front of the first stage of the compressor in a gas turbine engine to direct air into the first stage of the compressor at the proper angle for most efficient compression.

jam acceleration—Rapid movement of the throttle of a gas turbine engine that calls for maximum acceleration of the engine.

jet pump—A fluid pump that produces a low pressure by moving fluid at a high velocity through a venturi.

jet silencer—A noise suppressor used with a gas turbine engine that converts certain low-frequency vibrations into higher-frequency vibrations that are more easily dissipated as they mix with the surrounding air.

jet wake—A stream of hot, high-velocity gases from the exhaust of a gas turbine engine.

jet nozzle—The exhaust nozzle of a jet engine.

Jet-A—A kerosene-type fuel widely used in civilian jets. This fuel has a flashpoint between 110°F and 150°F and freezes at –40°F.

Jet-A1—Fuel similar to Jet-A, except that it contains additives to decrease the freezing point to –58°F.

Jet-B—A gasoline-type fuel similar to military JP-4 used mostly in military jets. It has a flash point of 0°F and freezes at –76°F.

JP-4—The primary fuel used for military jet aircraft composed of approximately 65 percent gasoline and 35 percent kerosene-type distillates. Similar to commercial Jet-B.

JP-5—Equivalent of Jet-A, a highly refined kerosene jet fuel.

kerosene—A light, almost colorless hydrocarbon liquid obtained by fractional distillation of crude oil. Kerosene is used as fuel for lamps and stoves and is the basis for gas turbine engine fuels. It is also known as coal oil.

kHz—Abbreviation for kilohertz (1000 cycles per second).

kilo—The metric prefix meaning 1000.

kilogram—Metric unit for measuring mass. One kilogram is the mass of 1000 cubic centimeters of pure water.

kinetic energy—The energy associated with motion.

knot—A measure of speed equivalent to 1 nautical mile per hour, 1.151 statute miles per hour, 1.688 feet per second, 1.852 kilometers per hour, or 0.5144 meters per second.

Krueger flaps—Wing leading-edge flap used on high-performance aircraft.

laminar flow—A type of airflow over an airfoil in which the air passes over the surface in smooth layers with minimum turbulence. Laminar flow decreases the drag produced by an airfoil.

landing roll—The horizontal distance from the point of touchdown to the point where an aircraft comes to a complete stop.

large aircraft—FAA designation for any aircraft weighing over 12,500 pounds.

leading-edge flap—A type of aerodynamic control surface used on some aircraft to increase the lift of the wing at slow speeds.

leading-edge radius—The radius of the curvature given to the leading-edge shape; the term is also typically expressed as a percentage of the chord-line dimension.

LEMAC—Abbreviation for the leading edge of the mean aerodynamic chord (MAC).

life-limited part—A part with a service life that is limited to a specified number of hours or cycles.

lift—The net force produced by an airfoil perpendicular to the relative wind.

local Mach number—The Mach number of any isolated or localized flow of air over an aircraft structure.

low-bypass turbofan engine—A turbofan engine with approximately the mass airflow through the core as is moved by the fan.

low-pressure compressor—The forward compressor, N_1, in a two-spool gas turbine engine, driven by the last stages of a turbine.

Mach number—A speed measurement that is the ratio of the true airspeed of the aircraft to the speed of sound under the same atmospheric conditions.

Machmeter—A flight instrument that indicates the flight Mach number of an aircraft.

mass—The amount of matter in an object. The attraction between the mass of an object and the Earth's gravitational field is known as weight.

maximum-allowable zero-fuel weight—The maximum weight authorized for an aircraft that does not include the weight of the fuel.

maximum-chamber location—The location along the chord line of the airfoil of the maximum chamber typically expressed in percentage of the basic chord-line dimension.

maximum chamber—Maximum displacement of the mean chamber line from the chord line of the airfoil. This dimension is typically expressed as a percentage of the basic chord-line dimension.

maximum thickness location—The location along the chord line of the airfoil of the maximum thickness typically expressed in percentage of the basic chord-line dimension.

maximum thickness—Maximum distance between the top and bottom surfaces of an airfoil typically expressed as a percentage of the basic chord-line dimension.

mean aerodynamic chord (MAC)—The chord of an imaginary rectangular airfoil that has the same aerodynamic force vectors as an actual airfoil.

mean chamber line—A line connecting all of the points midway between the upper surface and the lower surface of the wing. This line is also referred to as the *meanline* or *midline*.

medium-bypass turbofan engine—A turbofan engine with a bypass ratio of between 2-to-1 and 3-to-1.

millibar—A unit of measure in the metric system equal to a pressure of 1000 dynes per square centimeter.

motoring—The process of rotating the compressor and turbine in a gas turbine engine with the starter for any purpose other than starting the engine.

mule—A hydraulic power supply that can be connected to an aircraft to supply fluid under pressure to the hydraulic system when the engines are not operating.

nautical mile—A measure of distance used in air and sea navigation equal to one minute of latitude along the earth's equator (also equal to 6076.115 feet).

nominal value—A stated value that might be different from the actual value.

normal shock wave—The compressibility wave that forms ahead of a blunt object moving through the air at the speed of sound. The wave forms perpendicular to the air approaching the object.

partial pressure—The percentage of the total pressure of a mixture of gases produced by each of the individual gases in the mixture.

phosphate-ester hydraulic fluid—A synthetic hydraulic fluid used in high-performance aircraft hydraulic systems. Phosphate-ester fluid is fire resistant.

positive-displacement pump—A type of fluid pump that moves a specific amount of fluid each time the pump is actuated. Gear pumps, gerotor pumps, and vane pumps are positive-displacement pumps.

pressure vessel—The portion of the structure of a pressurized aircraft that is sealed and pressurized while in flight.

pylon—The structure that holds an engine nacelle to the fuselage or wing of a jet-propelled aircraft.

race—A hardened and polished steel surface on which antifriction bearings roll.

radial-outflow compressor—The same as a centrifugal compressor.

radio altimeter—A type of absolute altimeter that measures the height of an aircraft above the terrain by using a pulse of radio-frequency energy, also called *radar altimeter.*

radio magnetic indicator—An electronic navigation instrument that combines a gyro-stabilized magnetic compass card with an automatic direction finder.

ram recovery—The slight increase in thrust produced by a gas turbine engine caused by the increase in air density as the air is rammed into the inlet duct by the forward movement of the aircraft.

reaction engine—A type of engine that develops thrust by accelerating a mass of air inside the engine and forcing this high-velocity gas out through its tail pipe. Turbojet engines and rocket engines are reaction engines.

relative wind—The speed and direction of the air impinging upon a body passing through it. It is equal and opposite in direction to the flight path.

Reynolds number—A dimensionless number used to determine the flow characteristics of a fluid passing over a body. Reynolds number is determined by the velocity of the fluid, the distance the fluid travels over the body, and the kinematic viscosity of the fluid.

rotor—The moving part of an axial-flow compressor.

scavenger pump—A pump used in dry-sump lubrication systems that picks up the oil after it passes through the engine and returns it to the oil tank.

secondary air—Air that passes through a gas turbine engine compressor and is used for cooling rather than in the combustion process.

secondary fuel—The main flow of fuel from the fuel nozzles in a gas turbine engine combustor that allows the engine to produce its maximum power.

sequence valve—A type of flow-control valve that allows one system to fully actuate or operate before another system begins operation. Sequence valves are found in hydraulic systems to cause gear doors to open fully before the landing gear is released from uplocks.

simplex fuel nozzle—A type of fuel nozzle in which all the fuel is fed to the nozzle through a single fuel manifold.

single-spool engine—A gas turbine engine that has only one rotating mass.

slat—A secondary flight control on an airplane that allows it to fly at a high angle of attack without stalling. A slat is composed of a section of the leading edge of a wing that is mounted on curved tracks; the slat moves in and out of the wing on rollers. At high angles of attack, aerodynamic forces pull the slat forward on its tracks.

slot—A fixed slat on the leading edge of a wing.

stagnation point—The point at the leading edge of an airfoil where the air splits with a parcel of air passing over the airfoil and a parcel of air passing underneath it.

stall strip—Small triangular strip installed along the leading edge of the wing near the wing root to cause the wing root to stall before the tips, which increases controllability.

starter-generator—A combination electrical device that acts as a motor to start a small gas turbine engine and then is switched electrically to function as a generator.

static pressure—The pressure of a fluid that is not moving and measured perpendicular to the surface exposed to the fluid.

static thrust—The amount of thrust produced by a gas turbine engine when the engine is not moving through the air.

stator—The stationary portion of an axial-flow compressor in a gas turbine engine used to slow the airflow and increase its pressure.

stoichiometric mixture—The ratio of fuel and air in a combustion chamber that when burned leaves no uncombined oxygen or free carbon.

streamline flow—A liquid flow in which there is no turbulence.

subsonic flight—Flight in which all air passing over all parts of the aircraft is moving at a speed slower than the speed of sound.

symmetrical airfoil—An airfoil that has the same shape on both sides of its centerline.

tachometer generator—A small alternating-current generator mounted on and driven from the tachometer drive pad of an engine. The frequency of the ac power produced by the generator is proportional to the speed of the engine.

tail load—The aerodynamic force produced by the tail of an airplane. This force usually acts in a downward direction. The lift produced by the wing must overcome the weight of the airplane plus the downward-acting tail load.

tarmac—A word derived from tarmacadam, a mixture of tar and crushed stone often used as a paving material. The British use this term for the hard-surfaced area of an airport where aircraft are tied down and serviced.

thermal runaway—A serious condition in a nickel-cadmium battery in which the battery overheats due to excess current. Nickel-cadmium (nicad) batteries have very low internal resistance, but with an excessive current draw, even this low resistance will cause the battery to heat up. The center cells get hotter than the outer cells, which lose some of their heat to the atmosphere. When the cells overheat, their voltage and resistance both drop. As the generator starts recharging the battery, the center cells take most of the charge and get even hotter. This causes their voltage and resistance to drop, and they take more current. This condition continues until the battery is destroyed.

thrust horsepower—The horsepower equivalent of the thrust produced by a turbojet engine. Thrust horsepower can be determined by multiplying the net thrust in pounds by the velocity of the aircraft in feet per second and dividing the result by 550.

thrust-specific fuel consumption—A measure of the number of pounds of fuel burned per hour for each pound of thrust produced. A measure of fuel efficiency.

thrust—The aerodynamic force produced by a turbojet engine as it forces a mass of air to the rear.

time in service—The time from the moment an aircraft leaves the surface of the Earth until it touches down again at the next point of landing.

ton of refrigeration—A measure of the cooling effect of an air-conditioning system. The amount of cooling produced when 1 ton of ice melts in a 24-hour period.

total pressure—The pressure that a column of moving fluid would produce if it were stopped. The sum of the static and dynamic pressure caused by the impact of a moving fluid.

total temperature—The temperature of a fluid that is stopped from its motion. A sum of the static temperature of the fluid and the temperature caused by ram effect as the fluid is stopped.

transonic flight—Flight at speeds between approximately Mach 0.7 and 1.2.

triple-spool engine—A type of turbofan engine in which the fan section of the compressor is the first stage or N_1. The normal low-pressure compressor is N_2 and the high-pressure compressor is N_3. Each stage has its own turbine and rotates at its own best speed.

turbine—A wheel fitted with vanes or airfoils. Fluid flowing through the turbines produces a reactive force that causes the wheel to spin.

Ty-rap—Patented nylon strips used to tie bundles of wires together (also used to tie an arrested-person's hands together in lieu of handcuffs).

vector—A quantity that has both direction and magnitude.

velocity—A vector quantity expressing both the speed an object is moving and the direction in which it is moving.

venturi—A restriction in a tube designed to speed up the flow of fluid passing through it.

vertigo—A type of spatial disorientation caused by the physical senses sending conflicting signals to the brain.

very low frequency (VLF)—The range of radio frequencies between 3 kilohertz (3000 cycles per second) and 30 kilohertz (30,000 cycles per second) used for operations that require signals to travel great distances.

viscosity—The resistance of a fluid to flow. A measure of internal friction of the fluid.

vortex—A whirling motion in a fluid.

vortex generators—Small airfoils on the surface of and perpendicular to another airfoil. These are mounted in pairs and pull high-energy air down to the surface of the airfoil preventing air separation from the surface.

wing loading—The amount of load each square foot of the wing area must support.

zero-fuel weight—The weight of an aircraft with all of its useful load except fuel onboard.

Index

Illustration page numbers are in **boldface**.

ABOUT THE AUTHOR

Linda D. Pendleton has been an FAA-designated pilot examiner for airline transport pilot certificates and Cessna Citation type ratings since 1988. She worked for more than three years at FlightSafety International as a flight, ground, and simulator instructor in the Citation program. Pendleton wrote the FAA-approved flight, ground, and simulator instruction programs for the Cessna Citation type rating at FlightSafety's Long Beach, California, Learning Center. She also wrote the instructor's syllabi and trained instructors for the Citation courses.

She flies Citations and Lears as an FAR 135 captain for Chrysler Aviation in Van Nuys, California.

Pendleton is a Gold Seal Flight Instructor and has given more than 4000 hours of Citation instruction. She has been a professional pilot and flight instructor since 1975 when she started her career as a copilot on a turbine Beech 18 flying contract United States mail. Pendleton has flown for cargo, on-demand charter, and corporate operations. She was the chief pilot and director of operations for a small commuter airline in Chicago.